After graduating from the University of
Nottingham, Dr. Peel joined the staff of the
Botany Department of the University of
Hull in 1959 as an Assistant Lecturer. During
his stay at Hull he has carried out work on
the phloem physiology of higher plants,
using as one of his basic tools, the aphid
stylet technique. The topics covered by his
work include phloem movement and water
relations, metabolic studies using low temperatures and chemical inhibitors, ATP
relationships with transport, the mechanisms
involved in solute loading, growth regulator
and pesticide movement and growth
regulator effects upon nutrient transport.
This work has received considerable
financial support from several bodies,
particularly the Science Research Council.
Dr. Peel was appointed to a Readership in
Botany in 1971. He is married with two
young children.

Contents

Preface

The transport processes of plants present the investigator with some of the most fascinating and complex problems of any area of plant physiology, yet they have not generally received the attention they deserve as indispensable aspects of the life of higher plants. Although studies on nutrient transport can be said to have begun in the eighteenth century with the work of Stephen Hales (incidentally, his book *Vegetable Staticks*, reprinted in 1961, should be read by every serious student of translocation), progress over the past 200 years has been sporadic. Apart from a peak of activity during the late 1920s and early 1930s, produced mainly as a result of the efforts of Mason and Maskell, the only other notable period in the study of translocation has taken place during the last 15 years or so.

Undoubtedly, this latest upsurge of interest has been generated by the discovery of new techniques or the elaboration of old ones to probe the mysteries of the cells, buried deep within other tissues, which conduct solutes throughout the plant. Foremost among these techniques are radioactive tracers, incision or tapping methods for sampling the nutrient streams, and electron microscopy.

Prospective workers on translocation do not need to be physiologists. The field is wide open to those with aspirations in biochemistry (relatively little is known about metabolic processes in phloem and the other vascular tissues) or electron or optical microscopy, and to those who are just interested in plants. If the impetus of the last few years can be maintained, it is possible to visualise the unravelling of some of the major problems within the foreseeable future.

Transport of solutes in plants can take place over both short distances (measured in microns) and long distances (measured in centimetres or metres). Although both are equally important to the growth of all plants consisting of more than a few cells, it is the long distance systems, typified by the xylem and phloem of higher plants, which in many ways are the most fascinating, if only from the point of view of the distances involved.

The magnitude of the long distance transport processes can readily be appreciated in tree species. Here is a growth habit in which the two major sites of nutrient absorption and synthesis—the roots and leaves, respectively—are situated at either end of a long axis. The length of this axis in the extreme case of the Redwoods can be 100 m. Water and mineral salts must therefore move this distance from the roots to the topmost leaves, while assimilates from the leaves have to be transported, at least a considerable distance, from the lower branches to the roots.

Despite the obvious importance of these transport processes to the growth and development of land plants, we still lack a clear understanding of how nutrient movement is effected and controlled. This may be the reason why only a small amount of space is devoted to translocation in most textbooks on plant physiology. However, despite the deficiencies in our knowledge, there is a considerable body of data which has been accumulated over the past 40 or 50 years (and particularly during the past decade), which has not been adequately presented in book form at an undergraduate or first-year postgraduate level.

The greater part of this book deals with the physiology and cytology of phloem. This should not be taken as implying that the phloem is of greater importance than the xylem in the functioning of the plant. It is merely a reflection of the greater complexity of the phloem, in both structure and function, relative to the xylem. It must also be borne in mind that the xylem and phloem are not isolated from each other; not only are they anatomically close, but the two are also intimately connected in a physiological sense.

The emphasis of this book is upon function. Clearly, it is not possible to separate considerations of function from those of structure, but in view of the number of excellent publications on the gross anatomy of the vascular tissues, it does not seem necessary to duplicate them here. Discussion will therefore be largely confined to the severe problems associated with sieve element ultrastructure.

The first chapter deals with studies on the definition of the cellular pathways of transport; the second, with a consideration of how the mobility of solutes can be measured and the range of chemical species which are moved in xylem and phloem. Then follows a discussion on the concepts of velocity and rate.

The rest of the book is devoted to the characteristics of phloem transport and the ultrastructure of sieve elements, including such topics as the control of movement, solute-loading and -unloading mechanisms, the dependence of transport upon metabolic energy, bidirectional movement and water movement in phloem. Finally an account is given of the movement of endogenous growth regulators and a brief assessment of 'hormone-directed' transport.

Thanks are due to Drs. Behnke, Cronshaw, Gunning and Thaine, who have generously supplied the photographs illustrating aspects of the cytology of phloem; to those members of the Botany Department, University of Hull, who have given freely of their services in the preparation of the figures; to Miss E. Sharpe for typing the manuscript; and, finally, to my wife for her assistance in correcting the manuscripts and proofs.

Introduction

Although the main emphasis of this book will be laid upon xylem and phloem, it is impossible to deal with these two pathways in isolation; water and salts have to traverse the root cortex before they can enter the xylem, and sugars and other solutes elaborated in the cells of the leaf have to be moved into the sieve elements. In addition, as will be shown later, radial transport can readily occur between the phloem and xylem tissues.

In relation to the movement of solutes between the relatively non-specialised (in the sense of transport) parenchyma cells of the leaf, root and main axis, and between these cells and the specialised long distance transport conduits, probably the most important concept which has emerged is that of the symplast–apoplast system (Münch, 1930). This concept envisages the plant body as being capable of division into two distinct, though intimately connected, regions. One of these, the symplast, is composed of the whole mass of living material within the plant. Individual protoplasts of the cells are surrounded by the outer limiting membrane, the plasma-lemma. Vacuoles of the cells are limited by the tonoplast. The plasmalemma extends through the pits in the cell walls as the outer covering of the plasmodesmata, thereby extending as a continuous membrane which encloses the whole of the cytoplasm of the plant. In contrast to the symplast, the apoplast consists of the non-living cell walls of the plant. Like the symplast, the apoplast forms a continuous system throughout the plant which protects and contains the symplast, thereby giving form to the plant. *Figure Int.1* gives a representation of the symplast–apoplast systems.

This division of the plant into living and non-living systems necessarily means that solutes can move by quite different mechanisms in symplast and apoplast. In the latter, movement will be either diffusional or in a bulk flow of solution. Once a solute reaches the plasmalemma at the outer edge of the symplast, it is presented with a permeability barrier which it has to cross, either by the slow process of diffusion or, more generally, by an active transport mechanism.

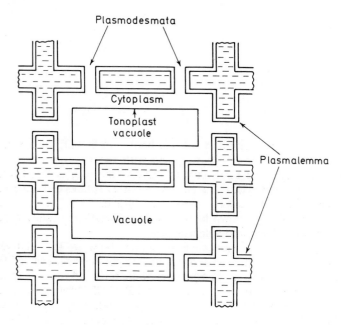

Figure Int.1 Diagrammatic representation of the symplast-apoplast system. Hatched areas show cell walls

Once across the plasmalemma, intracellular transport of solutes may be aided by cytoplasmic streaming. Intercellular movements in the symplastic system presumably occur mainly through the plasmodesmata. These movements are possibly largely diffusional, although it seems unlikely that movement through plasmodesmata is completely free and non-selective. Electron micrographs of plasmodesmata, e.g. those illustrated between ray parenchyma cells in Figure 5 of the paper by Evert and Murmanis (1965), show dark bands running transversely across the plasmodesmen which could indicate permeability barriers. Thus, active transport may be involved in the movement of certain solutes through plasmodesmata.

The results of some work (Hawker, 1965) on sugar cane have

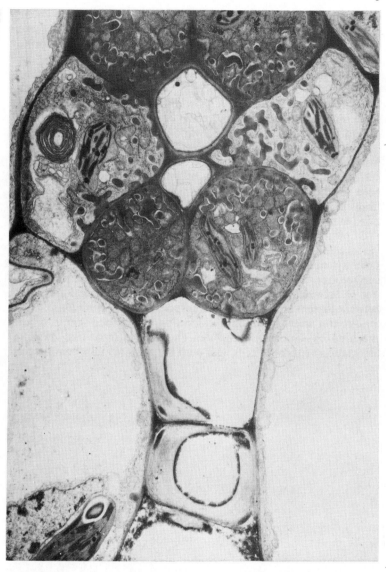

Figure Int.2 Electron micrograph of a leaf vein showing two types of transfer cells. Plasmodesmata occur between one type of transfer cell and the vascular cells of the vein. This type of transfer cell could therefore be involved in solute movement from mesophyll cells to sieve elements and in the interveinal recycling of xylem transported solutes. (From Gunning and Pate, 1974; reproduced by courtesy of the authors and McGraw-Hill)

indicated that the apoplastic cell walls may be involved in intercellular sugar transport rather than the plasmodesmata. Such a situation does not, however, appear to exist for sugar transport in willow phloem (Peel and Ford, 1968). Possibly, the relative importance of apoplast and plasmodesmata as intercellular transport systems differs according to species.

Before we leave the subject of short-distance intercellular transport, some observations on the occurrence of specialised parenchyma cells should be mentioned. Excellent work by Gunning and Pate (1969) has demonstrated that certain cells possess ingrowths of wall material and that, therefore, these cells have protoplasts with high surface-to-volume ratios. These cells, of which several types have been distinguished by electron microscopy, have been termed 'transfer cells', and are found in a wide variety of anatomical situations in most of the major groups of multicellular plants.

Of particular importance is the occurrence of transfer cells in the vascular tissues of higher plants. They are found associated with both xylem and phloem tissues, and in view of their enhanced surface area, are very possibly involved in the loading and unloading of solutes in xylem and phloem and in the radial transport between these two long-distance transport tissues. *Figure Int. 2* shows an electron micrograph of a transection of a minor vein of a leaf prepared by Gunning and Pate. This demonstrates the intimate association of transfer cells with both the phloem and xylem tissues.

1

The pathways of long distance transport

The movement of solutes from the roots to the leaves

Transport across the root cortex from the soil solution to the xylem vessels

Studies on the radial movement of solutes from the external milieu to the xylem vessels have been mainly concerned with ions. This is to be expected, since the bulk of solutes which are transported across the root system, at least under natural conditions, are ions on their way to the leaves and other aerial portions of the plant. Movement of ions across the root cortex has been the subject of study by a considerable number of investigators for a period of at least 50 years; even so, problems remain and there is no complete agreement on the pathways and mechanisms involved.

Before commencing a detailed analysis of the work which has been performed on the centripetal ion transport system of roots, it seems desirable to define the problems which have to be solved. Fundamentally, these problems can be stated in the following way: either ion movement across the root cortex is mediated through the symplast system with its attendant active, energy-requiring processes, or transport may be largely passive through the apoplast, ions being swept along in the water stream of transpiring plants. As we shall see, it appears possible that both pathways participate in transport, but the extent to which each contributes to the total ion flux could be dependent upon a large number of factors, two of the most important being the rate of transpiration and the 'salt status' of the plant. *Figure 1.1* illustrates the possible pathways and mechanisms involved in the ion transport system.

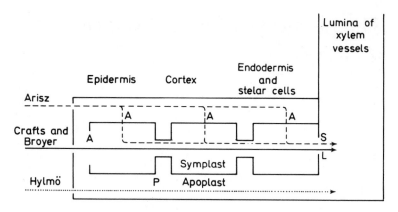

Figure 1.1 Schematic representation of possible pathways of, and processes concerned with, centripetal ion movement across root cortex. A, accumulation phase; S, secretion phase; L, leakage phase; P, passive apoplastic transport in the transpiration stream

The view of ion transport which has received most support is the symplastic theory, typified in all its essentials by the ideas of Crafts and Broyer (1938). These workers suggested that ions are taken up by the epidermis and cortical cells in a part of the root where accumulation processes are working at maximum efficiency. The ions are then moved through the continuum of the symplast by either diffusion or protoplasmic streaming, or both, until they arrive at the stelar cylinder. In the stele it is envisaged that the cells are unable to retain ions effectively because their accumulation capacity is reduced owing to a deficiency of oxygen. Thus the ions leak into the apoplast of the stele and thence into the lumina of the xylem vessels. Leakage of ions back into the cortex and the medium surrounding the roots is prevented by the suberised walls of the Casparian strip in the endodermal cell walls. This theory of symplastic ion transport has been mainly criticised from the standpoint that oxygen may not indeed be deficient within the stelar tissues.

Most workers, however, favour the concept of an active process being an integral part of the centripetal ion transport system. Either this active component is sited in the epidermis and cortex, as in the mechanism proposed by Crafts and Broyer (1938), or the movement across the cortex may be passive, the energy-requiring step taking place during the transport of ions from the apoplast of the cortex into the symplast of the stelar cells, thence into the xylem vessels (Arisz, 1945).

A completely opposite view of ion transport has been proposed by Hylmö (1953, 1958). This worker believes that ions are carried passively across the root in the transpiration stream. As we shall see later, there is considerable evidence that ion movement across the root can be affected by changes in the water flux through the root system.

Region of ion entry into the roots

The part of the root through which ions can enter on their way to the xylem is apparently limited to the portion which lies a short distance behind the apex. Older parts of the roots tend to undergo suberisation, which leads to the formation of a barrier which is relatively impermeable to water and ions.

Experiments by Wiebe and Kramer (1954), in which they exposed sequential lengths of intact barley roots to labelled ions, demonstrated that the greatest uptake activity was present some 3 cm behind the apex. The apices accumulated ions but transported only small quantities, the portion of the root lying between 1 and 6 cm behind the apex showing both accumulation and transport. In a further series of experiments Canning and Kramer (1958), employing labelled phosphates on corn, cotton and pea seedling roots, found a similar situation to that in barley, although in cotton and pea roots some translocation took place from the apex itself.

These results demonstrate that ions move into the xylem over a region several centimetres behind the apex, where the root is unsuberised and where the xylem elements have reached maturity.

Exudation phenomena in detopped root systems

Investigations on ion movement across the root cortex to the xylem have been performed on several types of systems, viz. excised roots, intact plants and detopped root systems. The latter system has proved to be very popular, since a flow of a dilute solution of salts can be obtained from the cut stump of the stem, and it has been shown that this fluid emanates from the xylem elements. Using such a system, it is possible, by changing the composition of the solution bathing the roots, to alter the flux rate and composition of the exudate, and from the data so obtained to draw conclusions concerning the processes which occur between the two ends of the system. Herbaceous plants which have been used in studies on the exudation process include sunflower, corn, tomato and castor oil.

Before we proceed with a detailed account of the exudation process, it seems desirable to add a note of caution. In common with many systems in botanical research, the exudation process is artificial; plants do not normally grow in a detopped condition, but possess aerial parts which lose water by transpiration. Thus, for a considerable part of their existence, the roots of intact plants will carry a transpirational flux of water which could affect ion transport across the cortex. Any conclusions which are drawn from excised root systems may therefore only present, at most, part of the picture. As will be mentioned later, certain workers have attempted to simulate transpirational water fluxes in detopped root systems by applying pressure to the solution bathing the roots.

The first point to be made about root exudation is that the movement of ions into the exudate must presumably involve an active process, at least in the absence of a transpirational flux of water. It is often found that the exudate contains ions, particularly potassium, at a higher concentration than that in the external medium. Of course, a difference in the concentration of an ion on two sides of a membrane system does not in itself necessarily mean that active transport of the particular ion is taking place. In the case of ions, a demonstration of a difference in electrochemical potential is necessary (Dainty, 1962).

An attempt to measure gradients of electrochemical potential across detopped root systems of *Ricinus communis* has been made by Bowling and Spanswick (1964). These workers measured the electrical potential difference between the exudate and the external solution, together with the concentrations of potassium and chloride in the exudate. Typical values for the former lay in the range −49 to −69 mV. Comparisons were then made between the Nernst potentials of the two ions, calculated from their concentrations in the exudate and external solution, and the observed potentials. With potassium the difference in electrochemical potential between the two ends of the system was small in a positive sense, or even negative. Thus, since the potassium ion is positively charged, this means it is probably moving passively across the root system. On the other hand, the electrochemical potential difference for chloride was large and negative, which would tend to move the negatively charged chloride ion into the external solution. Since there is a net flux of KCl into the sap, it must have been the case that the chloride ion was being actively transported, the potassium ion following passively to maintain electrical neutrality.

Some further evidence that the transport process involves active stages is found in the observation that the composition of the root exudate can be markedly different from that of the solution surrounding the roots. It is also evident that the uptake of ions from the external solution and their movement to the xylem vessels are to a degree separable processes, for exudation will continue for a considerable time after the roots have been placed in distilled water. Under these conditions ions which have been previously accumulated by the root system are transferred into the xylem. Van Andel (1953) has termed this process 'tissue exudation'.

The literature on root exudation is extensive. Early investigators concluded that the root system acted as an osmometer, a passive flux of water taking place across the system in response to a gradient across the root maintained by an active centripetal ion transport.

A contrary view of the process was put forward by van Overbeek (1942), as a result of his experiments carried out on root exudation from tomato. This worker argued that if exudation is purely an osmotic process, then, if the osmotic gradient across the roots is reduced to zero, exudation should cease. To test this possibility, van Overbeek placed detopped tomato plants into water and then added mannitol solutions until exudation ceased. He then returned the plants to water, when exudation restarted, and measured the osmotic potential of the exudate. He found that the osmotic potential of the exudate was considerably lower than the osmotic potential of the solution required to stop exudation, and suggested that water is moved by an active process across the root system. Cyanide was found to partially inhibit exudation, the extent of the inhibition being a measure, according to van Overbeek, of the water flux due to an active process.

The underlying fallacies in van Overbeek's work were exposed in the classic experiments of Arisz, Helder and van Nie (1951). These workers demonstrated that if the solution around the roots was changed from Hoagland's solution to one of higher osmotic potential (Hoagland's plus mannitol), the rate of exudation fell rapidly and then recovered to a value less than the rate in Hoagland's alone. The effect on the rate of exudation was reversible if the roots were returned to Hoagland's solution (*Figure 1.2*). By plotting the reduced rate of exudation, measured 30 s after transferring the roots from Hoagland's to Hoagland's plus mannitol, against the osmotic potential of the solution, Arisz *et al.* obtained a linear relationship indicative of an osmotic process (their Figure 4).

This inference was considerably strengthened by the demonstration that, on changing from a solution with a low osmotic potential to one of higher osmotic potential, the osmotic potential of the exudate did not stay constant but rose, thereby maintaining an

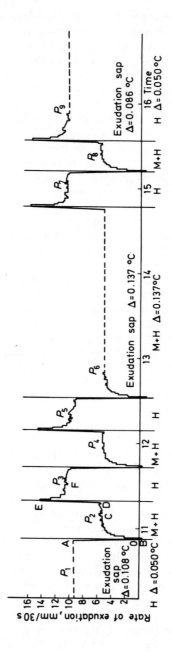

Figure 1.2 Influence of changing the osmotic potential of the medium surrounding the roots on the rate of exudation from a detopped tomato plant. The root system of the plant was continuously in a Hoagland solution (H), the osmotic value being increased by addition of mannitol (M + H). (From Arisz, Helder and van Nie, 1951, courtesy of The Clarendon Press)

osmotic gradient across the root system. For example, in Hoagland's solution with a freezing point depression of $-0.050°C$, the exudate had a freezing point in the range -0.108 to $-0.086°C$. In Hoagland's plus mannitol (freezing point, $-0.137°C$), the freezing point of the exudate dropped to $-0.184°C$.

The essential features of root exudation, according to Arisz *et al.*, can be summed up in diagrammatic form (*Figure 1.3*). With the root system at equilibrium with the surrounding medium, an active salt flux (S) proceeds from the root cells into the xylem. Owing to this flux, an osmotic potential gradient is maintained across the root; therefore a passive water flux (X) occurs in response to this gradient, the rate of X being equal to $k(OP_i - OP_e)$, where OP_i = osmotic potential of exudate, OP_e = osmotic potential of medium and k = water permeability of root. If OP_e is changed to a higher value, the gradient is immediately reduced and the water flux declines, but salt secretion still proceeds. This leads to an increase in concentration of the exudate (i.e. OP_i increases), and the system soon reaches a new equilibrium with a lower value of X.

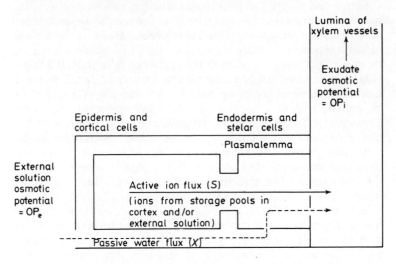

Figure 1.3 The root system as a dynamic osmometer. An osmotic potential gradient is maintained across the root cortex by an active flux of ions (S). Water follows passively in response to the gradient

Arisz and his colleagues concluded that both the water permeability of the protoplasm of the root and the active salt flux into the xylem could be influenced by the osmotic potential and the presence of ions in the external medium.

Concentration gradients across the root cortex and the 'leakiness'
of the stelar tissues

The hypothesis that symplastic transport of ions takes place across the root cortex, followed by leakage into the xylem vessels as proposed by Crafts and Broyer (1938), requires that two conditions be fulfilled. Firstly, there should be a gradient of ionic concentration across the cortex from epidermis to stele in order that diffusional movement may take place. Secondly, it is necessary that the stelar cells 'leak' ions rather than accumulate them.

During the past 10 years a number of workers have attempted to determine whether these prerequisites of the symplast hypothesis are fulfilled within a translocating root system. Two techniques have been used in an attempt to determine the changes which occur in the concentration of an ion from epidermis to xylem vessels during uptake by the root—histoautoradiography and X-ray microanalysis. The results from experiments utilising these techniques have shown a rather disturbing degree of variability.

Luttge and Weigl (1962) proposed that the endodermis was the site of active transport into the stele from the results of experiments with $^{35}SO_4$ and ^{45}Ca, using corn and pea roots. However, in certain of their experiments there were indications that accumulation of these labelled ions took place in the epidermal cells. S. F. Biddulph (1967) showed that labelled sulphate was apparently distributed in a uniform manner in absorbing bean roots, although ^{45}Ca tended to be accumulated to a greater degree in the epidermis and stele. Crossett (1967), using labelled phosphate on corn roots, found a higher concentration of this ion in the xylem vessels than in the stelar parenchyma and phloem tissues. He observed a gradient of the tracer across the root cortex with the highest concentration in the endodermis. The epidermis tended to have a higher concentration of ^{32}P than the adjacent cortical cells.

In some ways $^{32}PO_4$ is a difficult ion to use in transport studies: it is readily metabolised in plant tissues and soon becomes incorporated in a wide range of organic compounds, sugar phosphates, nucleotides and nucleic acids. It is very probable, therefore, that data from $^{32}PO_4$ experiments may be distorted by incorporation of the label into compounds not mobile in the system under investigation.

Lauchli (1967) has employed an X-ray microanalyser to determine phosphate, strontium and calcium distribution in corn roots. With phosphate the concentrations were highest in the epidermal regions and the phloem of the stele. The data for strontium and calcium indicated a higher concentration in the epidermis, with a uniform distribution in the remaining cross-section. Lauchli concluded from his results that the barrier to ion transport was located in the plasmalemma of the outermost cortical layer.

More recently, attempts have been made to measure electrochemical potential gradients of potassium across corn roots using microelectrodes inserted at different depths in the root tissues (Dunlop and Bowling, 1971). As pointed out previously, electrochemical potential gradients are more pertinent to considerations of ion movement than are mere concentration gradients. The results of the work of Dunlop and Bowling showed that the ability to accumulate potassium actively was the same for all the cells investigated in the root tissue. Consequently, there appeared to be no detectable gradient driving potassium across the root (*Figure 1.4*).

Dunlop and Bowling consider that their results are consistent with the symplasm hypothesis of movement, the lack of gradient being explained by reference to some calculations of Tyree (1970), who suggested that only very small gradients are required (about 1 mM across the whole cortex). Their data are certainly not consistent with the idea that the endodermis has a special secretory role, and there is no evidence from their experiments to support the concept that the stelar cells are more 'leaky' than those of the cortex.

However, work has been reported (Laties and Budd, 1964) which seems to demonstrate that the stelar cells can 'leak' ions. Experiments with decorticated steles of corn roots showed that freshly prepared steles rapidly leaked potassium and chloride ions. Luttge and Laties (1967) worked on the accumulation characteristics of ^{36}Cl and ^{86}Rb in intact roots, isolated cortical tissues and steles. Isolated steles absorbed ions at a very low rate immediately after isolation (although their accumulation capacity increased with ageing), which gives credence to the idea of 'leaky' steles. On the other hand, Yu and Kramer (1969) showed that steles isolated after a period in KCl solution contained concentrations of potassium and chloride at least equal to those present in the cortex. This was found even though the steles were washed for 30 min after isolation, which, according to the data of Laties and Budd (1964), should have removed 60% of the absorbed ions from the steles.

Ion transport through root systems in relation to water flux

So far in the consideration of centripetal ion transport through roots, we have been concerned with experimental results obtained in the main with excised roots, roots attached to seedling plants, or detopped root systems. Inevitably, this means that these systems are transporting ions with a relatively small or non-existent concomitant water flux. Even with exuding, detopped root systems, where a water movement across the cortex is patently occurring, the movement is merely an osmotic phenomenon in response to an active ion transport process.

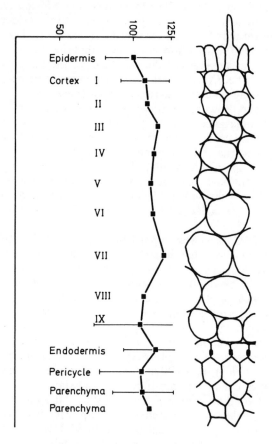

Figure 1.4 Vacuolar potassium activity of corn root cells bathed in 1.0 mM KCl *plus 0.1* mM CaCl₂. *Cortical cells are numbered from the outermost layer in roman measurements were made. (From Dunlop and Bowling, 1971, courtesy of The Clarendon Press)*

What we now have to consider is the situation in intact plants with expanded leaves, where transpiration can proceed rapidly, producing a rate of water flux through the root system many times greater than that in exuding root systems. It now seems to be well established that the rate at which ions are absorbed by the roots of intact plants and transferred to the leaves may be markedly affected by the rate of transpiration, at least under certain conditions. This does not, of course, necessarily mean that ions are being swept through the apoplast of the root directly into the xylem vessels (see *Figure 1.1*), although the operation of such a process cannot be dismissed. It could well be that high transpiration rates merely enhance the movement of ions to the aerial parts of the plant only after these ions have been symplastically transported across the root system. A further possibility is that the increase in the water potential gradient across the root system which occurs with high transpiration rates may indirectly affect ion transport by changing the permeability of the root system to ions. Let us now consider the evidence available, to see if it is possible to draw any conclusions as to the mechanisms involved in the effect of water flux on ion transport.

In 1943 Broyer and Hoagland published the results of some work on the effects of changing the water flux through the roots of barley plants on the uptake of potassium bromide. Their experiments were elaborated by dividing the plants into two groups, one of 'low salt status', the other of 'high salt status', the latter group having been grown in frequent changes of culture solution. They also arranged to alter the sugar status of the plants in the sense low salt : high sugar and high salt : low sugar. During the experimental period different batches of the low and high salt plants were subjected to environmental conditions which should have either reduced or enhanced transpiration. Some of the data from their experiments are shown in *Table 1.1*.

Table 1.1 INFLUENCE OF TRANSPIRATION ON ABSORPTION OF POTASSIUM AND BROMIDE IONS BY BARLEY PLANTS OF LOW OR HIGH SALT STATUS. (FROM BROYER AND HOAGLAND, 1943, courtesy of the *American Journal of Botany*)

Experimental conditions	*Water absorbed, ml/g fresh wt. of shoot*	*Salt absorbed, mequiv. $\times 10^2$ g total fresh wt.*		*Concentration in expressed sap of shoot, mequiv./l*	
		K	Br	K	Br
High salt : low humidity : light	8.10	5.20	6.07	216	58.0
High salt : high humidity : light	2.58	3.24	4.24	188	40.7
High salt : high humidity : dark	1.49	1.39	2.15	185	15.5
Low salt : low humidity : light	9.60	10.85	9.52	210	103.0
Low salt : high humidity : light	3.60	10.40	9.65	196	98.8
Low salt: high humidity : dark	2.52	8.75	9.13	178	75.5

These results show that under conditions of low humidity in light, the rate of transpiration was over three times greater than that found with high humidity in darkness, although the rate at which the potassium and bromide ions were absorbed was affected but little. On the other hand, with plants of high salt : low sugar status quite different results were found, the rate of ion uptake changing in the same sense as the rate of transpiration. Broyer and Hoagland concluded that the movement of ions across the roots of their plants was an active process and that the upward movement to the shoots took place in the xylem in a passive flow in the transpiration stream. In the plants of high salt status the latter process was considered to be limiting, and therefore variations in the transpirational water flux affected the rate of the whole process. Similar experiments by Russell and Shorrocks (1959) using ^{86}Rb and $^{32}PO_4$ gave essentially similar results to those of Broyer and Hoagland.

Other investigators, such as Brouwer (1954, 1965), have, however, found no positive correlation between transpirational water flux and ion uptake. Using *Vicia* seedlings, Brouwer (1954) utilised a micropotometer to obtain data on water and chloride uptake. Variations in water flux were not matched with variations in chloride uptake. Moreover, Brouwer demonstrated the independence of the two processes by application of metabolic inhibitors and changes in the osmotic potential of the solution bathing the roots; 10^{-5} M dinitrophenol inhibited chloride uptake but not water uptake, while the converse was found when the osmotic potential of the external solution was raised.

A completely opposite view of the pathway and mechanism of ion transport in intact transpiring plants has been given by Hylmö (1953). This worker observed that the absorption of calcium and chloride ions by pea seedlings varied with the rate of water loss from the leaves when the plants were subjected to conditions favouring, or depressing, transpiration. This is indeed similar to the results of Broyer and Hoagland, but Hylmö considered that the correlation was so close that he proposed a mass flow of solution through the apoplast of the root cortex as the major mechanism responsible for ion absorption and transport.

A criticism which has been levelled against Hylmö's conclusions is that in his experiments he did not investigate the effects of variation in the salt status of his experimental plants. If he had done this, he might have shown that the presence of a water/ion flux correlation was dependent upon salt status, as found by Broyer and Hoagland.

A possible difficulty with the concept of a passive, apoplastic mass flow of solution through the root cortex is to be found in the structure of the endodermis (Russell and Barber, 1960). This consists of a single layer of cells, the long axes lying parallel to the stele, with no intercellular spaces. The Casparian strip, consisting of a band of waxy substances, occurs on the cell walls (generally on the radial walls in gymnosperms and dicotyledons; on both the radial and tangential walls in monocotyledons). The Casparian strip, owing to its waxy nature, is considered to be impervious to water; therefore it can be argued that the apoplastic continuum of the root is broken at the endodermis.

On the other hand, evidence from plasmolysis forms in the endodermal cells indicates that plasmodesmata traverse the thickened walls. The cytoplasm of these cells remains firmly attached to the tangential walls after plasmolysis (Bryant, 1934). Thus, while the apoplastic system may be interrupted at the endodermis, the symplastic system appears to be continuous through this layer of cells, and this observation lends support to the concept of a wholly symplastic ion transport system across the root. Certainly, if the apoplast is broken at the site of the endodermis, it is difficult to envisage that a mass flow of solution could occur in a continuous stream between the external solution and the xylem vessels.

However, it is possible that the symplast/apoplast concept may be too simple in the context of ion movement in roots, and it is probably preferable to consider the situation in terms of 'apparent free space' (Briggs, 1957), this being the part of the tissue into which water and uncharged solute molecules can readily move by diffusion. (Owing to the presence of fixed negative charges within the apparent free space, anions are excluded from part of the free space. On the other hand, the concentration of cations in the free space can reach values greater than those in the external milieu, owing to exchange of protons associated with the fixed negative charges for cations, i.e. a Donnan exchange system exists within the free space.)

Some investigators (e.g. Dainty and Hope, 1959) consider that the apparent free space is mainly or entirely located in the cell walls and intercellular spaces. If this view is correct, then the apparent free space would lie outside the plasmalemma and would therefore correspond with the apoplast. An alternative suggestion (Hope and Stevens, 1952; Briggs, Hope and Pitman, 1958) is that the apparent free space may include some, at least, of the cytoplasm. If this latter view is correct, then an interruption of the root apoplast at the endodermis would not necessarily preclude a flux of solution through the cytoplasm of the endodermal cells in the apparent free space of the cytoplasm.

At all events, in some species of plants the endodermis does not possess a Casparian strip throughout its whole circumference. Certain cells of the endodermis known as passage cells can remain unthickened opposite the protoxylem elements. In such instances, therefore, no complete interruption of the apoplast occurs, and a free flow of water and ions through the walls of the passage cells can be readily envisaged.

An extension of the work on the relationships between water and solute fluxes was attempted by Jackson and Weatherley (1962). Instead of employing intact plants, these workers used detopped tomato and castor oil plants, changes in the rate of water flux through the root systems being induced by application of hydrostatic pressures to the external solution.

They found that the application of a pressure gradient across the root system produced at least a fourfold increase in the potassium flux into the xylem. This increase was independent of the presence of potassium in the external solution and it was therefore concluded that an effect upon 'tissue exudation' resulted from pressure application. Clearly, however, in the long term this effect must require enhanced uptake from the medium; otherwise the tissue stores would become depleted.

Jackson and Weatherley point out that their data do not mean that ions must necessarily be accumulated into a store before moving into the xylem, i.e. that storage is a prerequisite for the final stage of movement. It could well be that the ionic store is off the main pathway between external solution and xylem but is in dynamic equilibrium with ions on this pathway.

The main conclusions reached by Jackson and Weatherley were that pressure gradients had two main effects on potassium flux. The first of these was that pressure application produced a rise in the 'permeability' of the root tissue to potassium. This was inferred from experiments in which increments in the applied pressure were simultaneously balanced by increases in the osmotic potential of an external osmoticum, so that no change in the diffusion potential of the external solution was involved. Such experiments generally resulted in an increase in both water and potassium fluxes; and since the former was sometimes greater than the latter, the concentration of potassium in the exudate declined. However, in certain experiments the concentration as well as the flux of potassium rose, while the water flux remained constant. The second effect was termed a dilution enhancement of potassium flux. If the water flux into the vessels is increased by pressure application, the concentration in the exudate will fall; therefore the gradient for potassium between the store and the exudate will increase, which leads to an increased potassium flux into the xylem. This 'dilution' effect proved difficult to assess in terms of its relative importance with respect to the pressure-induced rise in potassium permeability.

Jackson and Weatherley concluded that the situation was complex and their work did not permit the formulation of a hypothesis which would cover all the known facts. However, they suggested that perhaps the most satisfactory generalisation lay in visualising the pathway between the medium and the solution as a catenary system. At one stage a movement of solution takes place; at another stage water and ions move independently of each other.

An elaboration of these ideas was made in a later publication by Bowling and Weatherley (1965). Experiments were performed on the relationship between transpiration and potassium uptake in castor oil plants. Although an increase in transpiration rate could lead to an enhancement of potassium uptake, a lag period was found between the two processes; in some cases potassium uptake did not increase until 7 h after the increase in transpiration rate (*Figure 1.5*). The presence of such a lag period would, according to Bowling and Weatherley, preclude the possibility of a direct pathway for the flow of a solution between the medium surrounding the roots and the xylem vessels.

Summing up the evidence, it would appear that the most probable pathway for ion movement across the root cortex involves at some stage an active, symplastic phase in both intact plants and isolated root systems. Whether this phase occurs early on in the transport process, i.e. at the epidermis or outer cortex as envisaged by Crafts and Broyer, is not known. It may occur further along the transport system at the endodermis; or, more likely, all the cells of the cortex are involved in the accumulation of ions from a solution which traverses the apoplast of the cortex. This could well lead to the situation in which the solution was completely devoid of ions by the time it reached the endodermis; thus only water might move the whole distance via a direct, apoplastic pathway.

The rate at which ions would be abstracted from the solution in the cortex would presumably depend upon a number of factors, one of which could be the 'ionic status' of the root cells. If this was high, accumulation rates would be low, and vice versa. It may be that a direct pathway for ion movement in a mass flow of solution does exist, but this can only be demonstrated, and indeed only becomes operative, in plants of high salt status. It could well be, even if it exists, of negligible importance, except in species growing in environments with high salt contents.

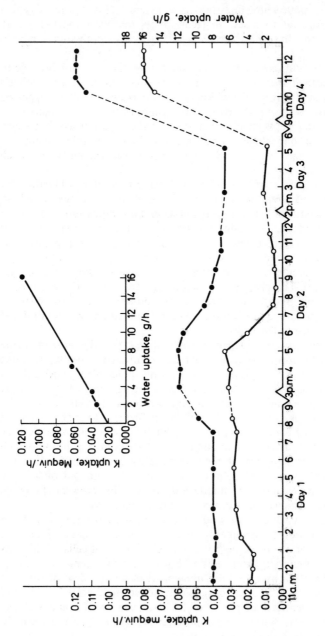

Figure 1.5 The trend in potassium uptake by an intact plant of Ricinus communis when the water uptake was altered in steps. Closed circles: potassium uptake. Open circles: water uptake. Inset: relationship between potassium uptake and water uptake. (From Bowling and Weatherley, 1965, courtesy of The Clarendon Press)

Transport along the main axis

Pathways of water transport in the transpiration stream

It is now generally agreed among physiologists that the xylem constitutes the channel for the transport of water in the transpiration
stream, although this conclusion has been questioned recently by
Wray and Richardson (1964). These workers applied tritiated
water to the roots of willow cuttings, maintained under conditions
which would promote rapid translocation. After absorption and
transport of the labelled water had proceeded for a given period,
the plants were frozen in solid CO_2 and the stems were cut transversely into a number of pieces. Tritium activity was assayed in the
inner and outer xylem of these stem pieces, and also in the extracambial tissues. Greater quantities of activity were found in the
outer than in the inner xylem, which confirmed the view which has
been held for a considerable time (Preston, 1952) that the transpiration stream moves largely in the outermost vessels. Rather surprisingly, however, Wray and Richardson showed that the activity
within the extracambial tissues was two or three times that in the
outer wood.

Of course, this does not necessarily imply that the extracambial
tissues are the pathway for the movement of the transpiration stream.
It probably means that centrifugal movement of the tritiated water
between the xylem and the outer tissues of the stem is very rapid.
In view of the weight of evidence in favour of the xylem being the
tissue concerned in the transport of water, it seems reasonable,
since no incontrovertible evidence exists to support other tissues in
this role, to conclude that the transpiration stream moves through
the xylem.

What is not certain is whether the pathway for water movement
in the xylem is confined to the lumina of the vessels or whether
considerable quantities of water can be transported through the
walls. Certainly, some evidence exists for the latter pathway playing
a not inconsiderable role, at least in certain tree species (Peel, 1965a).
Associated uncertainties arise when the transport of water across
the root cortex or leaf mesophyll is considered. Weatherley (1963)
has demonstrated that there are two pathways for water movement
in the leaf mesophyll: one with a high resistance to flow, possibly
located within the symplast, the other allowing water to move with
relative ease and probably located in the apoplast.

Quite recently, Crowdy and Tanton (1970) have attempted to
obtain visual evidence of the pathways of water movement in wheat
plants. This was done by applying lead-EDTA chelate to the roots

or to the leaves. After a period of time had been allowed for uptake, the lead was precipitated within the tissues by treating them with H_2S, location of the insoluble lead sulphide being determined by optical and electron microscopy.

Crowdy and Tanton concluded that bulk water movement took place in the lumina of the xylem vessels, since there was always a heavy lead sulphide deposit on the inner surface of the wall after root application of lead chelate. Outside the xylem, e.g. in the leaf mesophyll, the lead sulphide deposits were confined to the cell walls, being most dense in the region of the middle lamella. No deposits were found in the cells themselves. This seems good evidence for the view that bulk water transport in the leaf mesophyll is apoplastic.

It is not the purpose here to consider the mechanisms which have been proposed to account for the movement of the transpiration stream, although it now seems that most physiologists agree with the postulate of Dixon (1914) that transpiration from the leaves pulls water up through the plant along a gradient of negative pressure. However, there has been a considerable amount of controversy generated by the question as to whether columns of water under tension exist within the xylem of transpiring plants. Preston (1952) has produced evidence, both experimental and theoretical, against their existence, as has Greenidge (1957), employing a technique in which all the vessels of transpiring trees were opened to the atmosphere by means of double overlapping saw cuts. However, there is good evidence that high negative pressures can exist within the xylem of transpiring plants (Scholander *et al.*, 1966), and that water movement can occur through a system of microcapillaries when the xylem vessels are blocked by air (Scholander, Love and Kanwisher, 1955; Scholander, Ruud and Levestad, 1957; Peel, 1965a). Other writings on the subject of water movement in xylem to which the reader is referred are those of Zimmermann (1965), Slatyer (1967) and Zimmermann and Brown (1971).

Certainly, large quantities of water are moved in the transpiration stream at velocities up to 50 m/h. Moreover, the transpiration stream contains solutes, principally ions but also sugars and other organic solutes, albeit in low concentration (around 0.5% w/v).

Evidence for the presence of solutes in the xylem sap

Data indicating the presence of solutes within the xylem fluid come from two sources: composition of the sap which will exude from wounds made into the xylem of certain trees, or from the stems of

detopped herbaceous plants, and analyses of sap extracted from lengths of the stem of woody plants. In the following chapter we shall be looking at the mobilities of substances in the main long distance transport channels, but it would seem to be of value here to examine the validity of the methods employed to obtain xylem sap.

Exudate may be obtained by boring holes into the wood of trees in the early part of the year before the leaves have expanded. The species most studied in this way has been *Acer saccharum*, no doubt because of the ease with which sap may be obtained during the bleeding season. Other woody species which have been examined include *Betula*, *Vitis*, *Carpinus* and certain gymnosperms. It is usually assumed that the exudate comes from the lumina of severed xylem elements. Indeed, with certain herbaceous species Kramer (1949) was able to observe exudation from individual xylem elements.

The point at question, of course, is how close a relationship this bleeding sap has to the normal transpiration stream. A distinction has been made by Kramer (1949) between exudate produced by plants as a result of root pressure and that produced by local stem pressures. In detopped herbaceous plants and most woody species the exudation would undoubtedly appear to be the result of root pressure, and the solutes present should be representative, at least in a qualitative sense, of the nutrients being supplied by the roots to the aerial portions of the plant. In certain woody species, however, such as *Acer saccharum*, exudation can occur from detached pieces of stem, and it is possible that the exudate is produced as a result of local pressures which develop in the stem. Exudate from such species, therefore, may not truly represent the solute content of the transpiration stream. Moreover, it must also be borne in mind that even when the exudate is produced as a result of root action, its composition may differ from that of the xylem fluid when transpiration is taking place. Almost certainly, the exudates from detopped stems will be more concentrated than the transpiration stream, and it is conceivable that the relative masses of the various constituents may differ between the root exudate and the transpiration stream.

The other method used for obtaining samples of the xylem sap can be more generally applied, at least with woody species, since sap can be obtained at most times of the year. Short lengths of stem are cut from the plant and the sap is removed either by centrifugation, displacement by water pressure or the application of a vacuum to one end of the stem (Bollard, 1960). Using the latter technique, Bollard (1953) demonstrated that it is possible to remove up to 25% of the total moisture from the wood.

As with the exudation techniques, we have to examine the validity of these sap extraction procedures if we are to use them to determine the composition of the transpiration stream. Generally, it is believed that sap obtained from lengths of stem constitutes a representative sample of the contents of xylem vessels and is not seriously contaminated by the contents of other cells. A detailed investigation by Bollard (1953) has been carried out on apple. The removal of all the tissues external to the xylem before extraction of sap made no difference either to the quantity extracted or to the composition of the sap. Moreover, ethanolic extraction of the wood revealed that the extract contained relative amounts of nitrogenous compounds quite different from those present in the sap. Sugars were nearly always absent from the sap though they were present in large quantities in the ethanolic extracts. These results provide good evidence that the composition of the extracted sap is not markedly affected by the contents of living xylem cells.

Evidence that nutrients move in the transpiration stream

We now arrive at the point where two facts seem to be beyond dispute: the transpiration stream moves through the plant in the xylem, and this stream contains solutes such as ions and sometimes sugars. In view of these facts, it would seem to require only a simple extrapolation to reach the conclusion that solutes are transported from the roots to the leaves in the transpiration stream. Indeed, many physiologists did just this, but in fact it was found to be surprisingly difficult to prove, and several workers put forward what appeared to be cogent evidence against xylem movement of nutrients.

Probably the most vociferous opponent of xylem movement was Curtis. In a book published in 1935 he summarised the evidence for and against xylem movement of nutrients. In favour, he said, were the observations that the transpiration stream contained solutes and that this stream appeared to move in the xylem. Against this he put forward certain results he had obtained from girdled privet stems. Removal of the extracambial tissues over a short length of stem apparently resulted in the leaves of girdled branches receiving less nitrogen than the leaves of intact branches. From these data Curtis concluded that upward movement of nutrients took place largely in the extracambial tissues.

However, Mason and Maskell (1929) performed certain experiments on xylem translocation in cotton. Their experiments were short-term (over a period of hours), while those carried out by Curtis had lasted weeks or months, and it could have been that

with long-term experiments the ringing procedure might have affected water transport in the outer xylem vessels. Mason and Maskell demonstrated that a diurnal change occurred in the nitrogen content of cotton leaves, reaching a maximum concentration in the early evening. This diurnal fluctuation was observed in the leaves of both girdled and control plants, the magnitude of the change being statistically the same between the two groups. Thus, Mason and Maskell argued that girdling had no effect upon nitrogen transport into the leaves and therefore must take place in the xylem.

Further investigations on this problem had to wait until radio-tracers became available. It may be thought that such a powerful technique could readily provide an unequivocal answer to the problem, but in fact the first results were disappointing. However, failure was almost certainly due to a combination of long experimental periods and conditions under which the rate of transpiration was low. As we shall see in the final section of this chapter (pp. 35–40), radial movement between xylem and phloem can occur with extreme ease, and such movement almost certainly bedevilled the experiments of the earlier workers.

In 1939 Stout and Hoagland performed their classic work on the upward movement of labelled ions in willow and *Pelargonium*. They grew willow cuttings which possessed three branches, termed 1, 2 and 3. A 9 in slit was made longitudinally in the bark on opposite sides of branches 1 and 2 (*Figure 1.6a*). The bark was gently pulled away from the wood on each side and paraffined paper was inserted between the two tissues. The stripped areas were wrapped in paraffined paper to prevent loss of moisture. Branch 3 was left intact, except that paraffined paper was wrapped around the outside as with branches 1 and 2. After $1\frac{1}{2}$ h, the paper between the bark and wood on branch 2 was removed, the bark bound back, and the waxed paper replaced on the outside. $^{42}KNO_3$ was applied to the solution surrounding the roots and a fan was used to accelerate transpiration. After a total ^{42}K application period of 5 h, the experiment was terminated and the plant was sectioned for analysis in the manner shown in *Figure 1.6(b)*. The results of these analyses are shown in *Table 1.2*.

These data show that significant quantities of radioactive potassium had moved into all parts of branch 1, except the centrally located sections of the stripped bark. Wherever the bark was in contact with the wood, the amount of labelled potassium was as high in the bark as in the wood of the same section. The fact that potassium moved into the bark of the stripped portion of branch 2 was interpreted as meaning that negligible injury was caused to the bark by the stripping operation.

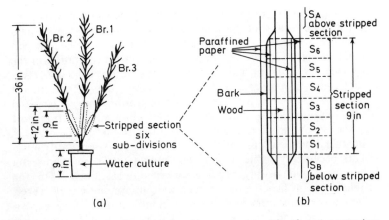

Figure 1.6 (a) Form of willow cutting used in experiments with radioactive potassium applied to root system. (b) Longitudinal section of stripped section of branch 1. (From Stout and Hoagland, 1939, courtesy of the American Journal of Botany)

Stout and Hoagland interpreted their results as meaning that, under conditions where the transpiration stream was moving rapidly through the xylem, salts absorbed by the roots were transported to the leaves in this flow of water. Rapid centrifugal transport to the bark could clearly occur where bark and wood were in contact. They also concluded that a rather slow, longitudinal transport could take place in the phloem; with long application periods, radioactive potassium could be found in all sections of the stripped bark of branch 1.

Table 1.2 MASS OF RADIOACTIVE POTASSIUM IN SECTIONS OF WILLOW STEM AFTER AN ABSORPTION PERIOD OF 5 h. (FROM STOUT AND HOAGLAND, 1939)

		Branch 1 Stripped $1\frac{1}{2}$ h before absorption period		Branch 2 Stripped $1\frac{1}{2}$ h; bark replaced before absorption period		Branch 3 Intact	
		p.p.m. bark	p.p.m. wood	p.p.m. bark	p.p.m. wood	p.p.m. bark	p.p.m. wood
Above strip S_A		53	47	78	52	64	56
	S_6	11.6	119	145	92		
	S_5	0.9	122	155	113		
Stripped	S_4	0.7	112	130	89	87	69
section	S_3	<0.3	98	132	91		
	S_2	<0.3	108	133	80		
	S_1	20	113	133	107		
Below strip S_B		84	58	137	73	74	67

Up to the present time, no one has seriously questioned the conclusions drawn by Stout and Hoagland, and their work still remains as definitive as when it was first performed.

The movement of solutes from the mature leaves to the roots and immature aerial parts of the plant

The characterisation of the sieve elements as the cells responsible for the transport of solutes from the mature leaves to the storage and growth regions of the plant has taken a very long time. In this respect, of course, it is typical of the whole field of phloem physiology; an idea or hypothesis suggested 50 or 100 years ago, although it receives support at the present time, may not have sufficient weight of evidence in its favour to be universally accepted. This (i.e. the interval of time between suggestion and acceptance) can be taken as a measure of the difficulties involved in probing a system composed of different cell types which function together as a unit. Thus, very frequently, the investigation of a complex cell system becomes dependent upon the elaboration of new techniques, many of which have their origins in other scientific disciplines.

The sieve tube was discovered by Hartig in 1837, and was immediately recognised by this worker as the probable channel for the movement of materials in the phloem. Incisions made into the phloem of trees were found to produce an exudate rich in carbohydrates, which Hartig presumed to emanate from the sieve tubes.

However, in the absence of direct evidence for the transport functions of the sieve tube, a number of workers attempted to show, by indirect means, that the sieve tubes could not possibly be the transport channels. Foremost among these workers were Dixon and Ball, who in 1922 published the results of their work on the movement of carbohydrates into the tuber of the potato. Their main objection to the sieve tube was a teleological argument based upon the structure of the sieve tube, viz. how can the sieve tube, consisting as it does of a file of sieve elements with frequent cross-walls (the sieve plates), even though these possess perforations (the sieve pores), possibly support a flow of a solution of sugars?

What Dixon and Ball attempted to show was that the velocity of flow of this hypothetical solution was too great to be within the sieve elements. Although their conclusions were subsequently proved to be wrong, their data remain a classic in the field. Since we shall be referring to their results later (Chapter 4), it seems appropriate at this point briefly to review their results.

Working on the assumption that all the carbohydrate of a potato tuber must have been translocated through the tissues of the stolon bearing the tuber, they argued as follows. The final weight of the tuber after 100 days was 210 g, of which 50 g was carbohydrate. The total cross-sectional area of the phloem was found to be 0.004 22 cm^2. Dixon and Ball then made the not unreasonable assumption that the carbohydrate had moved through the stolon as a 10% solution of sucrose. Therefore, to transport 50 g of carbohydrate into the tuber, some 500 cm^3 of this solution must have flowed through the stolon in 100 days, i.e. 5 cm^3/day. Now, if this 5 cm^3 of solution had been contained in the phloem, the length of phloem required would be 5/0.004 22, i.e. somewhat greater than 1000 cm. Thus the velocity of flow must have been at least 1000 cm/day or approximately 40 cm/h. This value, Dixon and Ball suggested, was vastly too great for the flow to be taking place in the phloem. Of course, this velocity would have to be increased several times if the hypothetical solution had been flowing only through the sieve tubes, for the cross-sectional area of the sieve tubes is somewhat less than that of the whole phloem.

The experiments of Mason and Maskell

Some work by Mason and Maskell on xylem transport was quoted above (p. 24). Although the investigations by these workers are now relatively ancient, no work on translocation would be complete without numerous references to them, such are the quality and quantity of the data which they produced. An excellent example of the enduring nature of their work is to be found in the investigations performed on the pathway of movement of organic solutes exported by the leaves of cotton. There is insufficient space here to describe their work on the transport pathway of organic solutes in detail, but it is hoped that the following summary will give a true impression of the scope and depth of their work on this problem.

Mason and Maskell (1928a) published the results of some work in which they demonstrated a diurnal fluctuation in the carbohydrate content of leaves. This fluctuation was mirrored 2 h later by a comparable change in the carbohydrate content of the bark situated 50 cm below the leaves (*Figure 1.7*). No marked variation took place in the carbohydrate content of the wood, however. Statistical analysis of the data showed a marked correlation between the changes in carbohydrate content of the leaves and the bark. Even better correlation was obtained if the values for the bark sugars were moved 2 h forward (this was reasonable because of the spatial separation of the leaves and the bark), and if sucrose values were used rather than total sugars (*Table 1.3*).

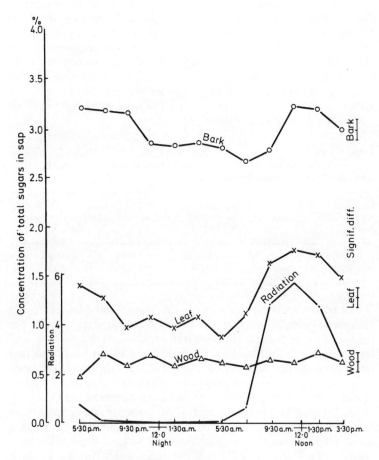

Figure 1.7 Concentration of total sugars in sap of leaf, bark and wood of cotton plants (g sugar per 100 cm³ *sap). 'Radiation' measured as the difference in* cm³ *between evaporation from black and white atmometers is also plotted. Values of the 'significant difference' for the sugar concentrations of leaf, bark and wood are shown on the right of the figures. (From Mason and Maskell, 1928a, courtesy of The Clarendon Press)*

Mason and Maskell took these data as cogent evidence for the movement of sugars, particularly in the form of sucrose, in the bark. However, realising that these experiments were not conclusive, they followed them with an exhaustive series of short-term ringing experiments. As a preliminary to these surgical experiments, they showed that ringing had no deleterious effects upon water transport in the xylem, at least over a period of hours, for no differences in the water contents of leaves from ringed and control plants could be detected. The variations in the surgical procedures they used were legion, but in summary it can be said that any treatment which interrupted the bark caused a cessation of sugar transport to the regions of the plant below the interruption. Complete girdling of a stem caused an increase in carbohydrate content above the girdle in both bark and wood tissues, and a fall in carbohydrate in these tissues below the girdle (*Figure 1.8*).

Table 1.3 COEFFICIENTS OF CORRELATION BETWEEN SUGAR CONCENTRATIONS IN LEAF, BARK AND WOOD OF COTTON PLANTS. (FROM MASON AND MASKELL, 1928a, courtesy of *The Clarendon Press*)

Correlation between leaf and:	Direct	Total sugars Bark values moved forward 2 h	Direct	Sucrose Bark values moved forward 2 h
Bark	0.76	0.94	0.94	0.96
Wood	0.01	0.45	0.26	0.02

The results of these experiments led Mason and Maskell to the inescapable conclusion that the bark tissues were responsible for the transport of sugars. However, the experiments did not give any information on the cells within the bark which constituted the actual transport conduits. Mason and Maskell (1928b) therefore attempted to obtain a more precise location of the transport channel. They had already demonstrated (*Table 1.3*), in their experiments on the correlation between leaf/bark and leaf/wood carbohydrate content, that the greatest degree of correlation was found when they utilised sucrose rather than total carbohydrate values. From this they argued that if they could locate the cells within the bark which contained the greatest concentrations of sucrose, then these cells would probably be the transport conduits.

Mason and Maskell divided the bark of cotton plants into inner, middle and outer regions, and then determined the percentage area of each region occupied by sieve tubes, ray and cortical tissues. They also analysed the three regions for sucrose content. From these data they were able to calculate that the sieve tubes contained the highest concentration of sucrose, the ray tissues less and the cortical tissues the least. Therefore they concluded that the sieve tubes were the channel of transport for carbohydrates.

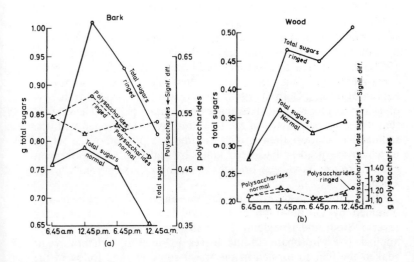

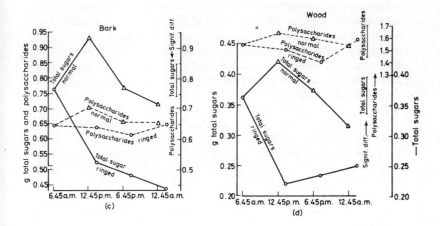

Figure 1.8 (a) Mass of total sugars and polysaccharides in the bark of a section above the ring. (b) Mass of total sugars and polysaccharides in the wood of a section above the ring. (c) Mass of total sugars and polysaccharides in the bark of a section below the ring. (d) Mass of total sugars and polysaccharides in the wood of a section below the ring. (From Mason and Maskell, 1928a, courtesy of The Clarendon Press)

Experiments with radioactive tracers on surgically treated plants

A continuation of the work of Mason and Maskell on the pathway of transport of solutes exported by the leaves was initiated in the 1940s, when radioactive isotopes had become available. As with the experiments on xylem transport, the first attempts gave somewhat equivocal results. Colwell (1942) investigated the basipetal transport of [^{32}P]phosphate applied to leaves of squash plants. He showed that if a large percentage of the total leaf area was flooded with the radioactive solution, then activity would pass a portion of the petiole in which the phloem had been killed by scalding. This movement must therefore have been brought about by reverse currents in the xylem.

Much more cogent evidence for movement in the phloem was obtained by O. Biddulph and Markle (1944), who performed a reverse 'Stout and Hoagland' experiment using cotton plants. They applied [^{32}P]phosphate to the leaves by incising a lateral vein; soaked the flap so formed in the tracer-solution; and followed the movement of activity down the stem, which had been surgically treated in a manner essentially similar to that shown in *Figure 1.6(b)*. Where the bark and wood had been separated throughout the experiment by waxed paper, virtually all the activity in the treated section of the stem was found in the bark. However, in experiments where the bark had been stripped and then bound back to the wood, detectable quantities of activity were found in the latter tissue. Biddulph and Markle concluded that transport of the tracer had taken place through the bark tissues, with centripetal migration to the wood where the tissues were not separated.

Before we leave these experiments, some elegant work by Rabideau and Burr (1945) deserves to be mentioned. Using bean plants, they applied $^{13}CO_2$ to the leaves and [^{32}P]phosphate to the roots. They were able to show that localised killing of the phloem by application of hot wax, completely stopped the upward or downward movement of ^{13}C-labelled assimilates from the leaves. However, this treatment did not affect the upward transport of ^{32}P-activity in the transpiration stream; therefore xylem transport was not affected by the heat treatment.

Exudation phenomena

It has been known for many years that incisions into the bark of certain trees will produce a flow of solution which contains a wide variety of solutes (Zimmermann, 1960b). Exudation may also be obtained from cuts made into the stems of herbaceous species, viz. cucurbits (Crafts, 1936; Moose, 1938) and *Yucca* inflorescence stalks (Tammes and van Die, 1964). This exudate is generally held to come from the sieve tubes and is thought to give a fairly good picture of the contents of functioning sieve tubes. We shall be returning many times to the subject of exudation processes, but, in the present context of the pathway of transport, the observations on exudation through the severed mouthparts of aphids are most valuable.

The use of aphids in the study of phloem physiology largely followed on from the observations of Kennedy and Mittler (1953). They found that individuals of *Tuberolachnus salignus* (Gmelin) could be anaesthetised while they were feeding on willow and the mouthparts severed, leaving the stylets embedded in the bark. From the cut end of the stylets drops of solution (*Figure 1.9*) exude at a rate of 1–4 µl/h. This solution was rich in sucrose and amino acids, and Kennedy and Mittler demonstrated that the tip of the stylets pierced a single sieve element.

Confirmation that aphid stylets enter sieve elements has been given by Zimmermann (1963), who has published an excellent picture of the tip of the stylet of *Longistigma caryae* (Harr) inserted in the sieve element of *Tilia americana*. Similar observations have been made by Evert *et al.* (1968), using the same species as Zimmermann, and by Kollmann and Dorr (1966), using a species of *Cupressobium* which feeds on juniper.

Exudation from severed stylets can proceed for hours or even days, during which time relatively large quantities of a number of nutrients are exuded. Weatherley, Peel and Hill (1959) calculated that an exudation rate of 1 µl/h from *Tuberolachnus* stylets involved emptying approximately 100 sieve elements per minute. Thus the sieve elements must be able to transport solutes rapidly towards the site of a stylet puncture.

Evidence from autoradiographic studies

S. F. Biddulph (1956) demonstrated that [^{32}P]phosphate or [^{35}S] sulphate, applied to the leaves, moved in the phloem. More recently, in an excellent piece of work by Fritz and Eschrich (1970), direct visual evidence has been obtained that it is the sieve tubes in the phloem which are the transport conduits. Application of ^{14}C-labelled phenylalanine or ^{14}CO$_2$ to the leaves of *Vicia* plants was followed by freeze drying, embedding and sectioning of segments of the stem below the application leaves. Exposure of the sections on X-ray film revealed that most of the activity was confined to the sieve elements, with very little present in either the companion or parenchyma cells.

Figure 1.9 Exudation of sieve tube sap from severed mouthparts of Tuberolachnus salignus *embedded in the bark of a willow stem*

The movement of solutes in radial and tangential transport systems

Circulation of nutrients within plants

In dealing with the pathways of solute transport within the plant, we are not only concerned with the long distance, longitudinal systems—the phloem and xylem—but also with the systems which move materials over relatively short distances. Clearly, it would be of little use to the plant to be able to transport materials from the roots to the leaves and vice versa, if it were not also able to convey nutrients to the cells of the stem which lie some little distance lateral to the main transport pathways. There is ample evidence in the literature for the presence of radial and tangential transport systems which move nutrients, not only from vascular to non-vascular tissues, but also between the vascular tissues themselves.

Much of the evidence for the functioning of radial and tangential systems has been obtained from experiments in which the circulation patterns of nutrients have been studied. As long ago as 1877, Hartig proposed that a circulation of carbohydrates might take place on an annual basis in trees. He concluded that assimilates moved from the leaves into the bark, where they were stored in the parenchyma and ray cells during the winter. In spring these assimilates were re-mobilised and moved into the xylem vessels, whence they ascended in the current of water. It has also been known for a considerable time that nitrogenous substances can be re-exported from senescing leaves to parts of the plant which are still growing.

The data of Mason and Maskell (1931) indicated that certain nutrients such as phosphorus and potassium could be circulated within the plant without an intervening storage period. More recently, O. Biddulph *et al.* (1958) have shown that some solutes can circulate freely. These workers applied radioactive phosphate, sulphate or calcium to the roots of red kidney bean plants for a period of 1 h, subsequently returning the plants to non-radioactive culture solution. At subsequent intervals, plants were harvested and the distribution of the tracer determined by whole plant autoradiography. Biddulph and his co-workers demonstrated that a portion of the labelled phosphate continued to circulate during the whole of the 96 h experimental period. The labelled sulphate was originally mobile, but this mobility was quickly stopped by retention in young leaves. The radioactive calcium did not circulate at all, remaining in the leaves after its delivery to these organs in the transpiration stream.

The radial transport system

The experiments on circulation of nutrients described above demonstate quite unequivocally that certain solutes can ascend the plant in the transpiration stream, then move from the xylem into the phloem and be transported downwards to the roots. We have already seen (p. 25) that the extent of centrifugal movement from the xylem to the extracambial tissues can be very considerable (*Table 1.2*, p. 26; data of Stout and Hoagland, 1939). Also, it is known that ions such as potassium and sodium may move centrifugally with apparent ease into the sieve elements (Peel, 1963).

Movement centripetally from the phloem to the xylem has been demonstrated by a number of workers. S. F. Biddulph (1956) provided excellent autoradiographic evidence that centripetal movement of radioactive sulphate and phosphate can occur in bean plants. Similarly, Webb and Gorham (1965a) have demonstrated centripetal transport of ^{14}C-labelled assimilates in squash plants. Peel (1967) has shown that labelled ions and sugars may move centripetally, not only into the living xylem cells but also into the xylem sap of segments of willow stem.

Having clearly established that radial movement occurs in plants, we now have to consider the characteristics of the transport systems, i.e. the amount of a solute which can be transported in a given time, the apparent velocity of movement (cf. Chapter 4) and the pathways which may be involved.

Let us first consider centrifugal transfer of solutes from the xylem to the phloem, since, in general, less information is available in the literature on movement in this direction. Apart from the data of Stout and Hoagland (1939), some unpublished data on the transfer of ^{14}C-labelled sugars and ^{32}P-labelled phosphates in willow stems have been produced by Peel and Ford (*Table 1.4*).

These results clearly show that centrifugal transport can take place in stems to a considerable degree. Indeed, in some of the experiments there was appreciably more ^{14}C-activity in the bark than in the wood at the termination of the experiment. It is also clear that considerable variation occurred between different stems in the magnitude of the centrifugal transport process. It should also be noticed that variability was shown in the extent of ^{14}C-labelled sugar movement in a centripetal direction (*Table 1.5*, p. 39). The cause of this variability is not known, for as yet we do not understand the factors which control the rate of radial movement of solutes. However, it is quite possible that future research will reveal a similar mechanism to that demonstrated with longitudinal movement in phloem, where sources and sinks play a major role (Chapter 5); in the experiments presented in *Tables 1.4* and *1.5* it is quite possible that the wood and bark had different source/sink capacities in the various stems employed.

Table 1.4 THE TRANSFER OF ^{14}C-LABELLED SUGARS AND ^{32}P-LABELLED PHOSPHATES BETWEEN THE WOOD AND THE BARK OF SEGMENTS OF WILLOW STEM (FROM PEEL AND FORD, UNPUBLISHED)

| Experiment | Ratio, activity in bark/activity in wood | |
	[14]C	[32]P
1	0.36	0.04
2	1.15	0.18
3	1.10	0.19
4	5.85	0.07
5	0.09	0.03

2μCi of [^{14}C]sucrose and of [^{32}P]phosphate applied via a hole drilled into the centre of the xylem. Experiments of 24 h duration.

As well as naturally occurring solutes such as sugars and phosphates, alien substances, e.g herbicides (Field and Peel, 1971a), can undergo extensive centrifugal transport in stems.

The problem which now has to be faced is the pathway taken by solutes moving from wood to bark, and a little thought will make it obvious that two possible pathways could exist. Since solutes traversing the xylem in the transpiration stream are already present within the apoplastic system of the plant, it could be that a diffusional movement might occur through the apoplast across the cambium into the apoplast of the bark. Also, it is possible that uptake could take place into the symplast of the xylem, followed by transport in the ray cells, where it would presumably be accelerated by protoplasmic streaming. The 'transfer cells' of Gunning and Pate (1969), found to be associated with the vascular tissues of certain species, may be involved in symplastic transport of solutes between xylem and phloem.

It is possible that both apoplastic and symplastic pathways exist. Certainly, with the somewhat artificial system constituted by isolated strips of willow bark, there is good evidence for the operation of a 'direct' apoplastic pathway and a symplastic 'indirect' pathway (Hoad and Peel, 1965b). This inference has been supported in the case of ions and sugars by the work of Ford (1967) and Peel and Ford (1968). In the first instance, Ford demonstrated that application of labelled ions to the cambial surface of bark strips for a short period of time, after which they were removed by washing, led to changes in the activity of sieve tube exudate, which fell into two phases. The first phase was considered to be produced by apoplastic transport, the second by symplastic movement. Peel and Ford demonstrated that the pattern of distribution of activity in sieve tube exudate was dependent upon whether ^{14}C-labelled sugars were being transported by a symplastic system or whether an apoplastic movement could have been concerned.

Of course, although there is good evidence for an apoplastic pathway in isolated bark strips, this does not mean that such a pathway is operative in intact stems. Stripping the bark disrupts the cambium, and it may be that this tissue forms an effective barrier to apoplastic transport in the undisturbed system.

When centripetal transport between phloem and xylem is considered, we find that work on this aspect of movement has received more attention than movement in the opposite direction. A number of determinations of the magnitude of this process may be found in the literature. O. Biddulph and Cory (1957) observed that with a 1 h application period the xylem of bean plants contained approximately 24% of the total ^{32}P- and ^{14}C-activities exported from the leaves to which these tracers were supplied. Comparable results were found with the cotton plant by O. Biddulph and Markle (1944).

In one of their classical researches on translocation in cotton, Mason and Maskell (1928b) were able to arrive at an estimate for the mass transfer of sugars between the ray cells and the xylem parenchyma. The values they obtained lay between 6.24×10^{-4} g and 11.41×10^{-4} g per square centimetre of ray area per hour for a concentration gradient of 1% hexose per centimetre. The rate of mass transfer of hexose by thermal diffusion, expressed in the same units, is 9×10^{-5}; thus the observed rates were an order of magnitude greater than could be accounted for by physical diffusion. However, Mason and Maskell considered that their measurements did not necessarily rule out movement by diffusion.

Webb and Gorham (1965a) made measurements on the apparent velocity of movement of ^{14}C-labelled assimilates from the phloem of squash plants. They allowed a leaf to photosynthesise in ^{14}CO$_2$ and then determined the distribution of the ^{14}C-activity in certain tissues by autoradiography. They concluded that the velocity of movement lay between 1 and 6 cm/h, depending upon the age of tissue concerned; the younger the tissue, the greater the velocity of movement. Webb and Gorham considered that transport was most probably taking place in living cells, mediated by cytoplasmic streaming, since the measured velocities of streaming in a range of cell types were of the same order of magnitude as the velocity of radial transport.

King (1971) has carried out a study on the transport of ^{14}C-labelled assimilates in leafy segments of willow. He not only concerned himself with movement between the bark and wood but also made measurements on the transport of the labelled sugars between the cells of the xylem and a flow of water through the xylem which simulated the transpiration stream. He demonstrated that the proportion of the [14C]sugars which moved into the xylem stream was independent of flow rate, an observation which led him to the conclusion that it was unlikely that the labelled sugars were moving from the bark into the xylem stream by an apoplastic pathway.

King performed his experiments by applying [14]CO$_2$ to a leafy shoot borne on a stem segment of willow. Xylem effluent was then collected over a 24 h period, after which time the bark and wood tissues were assayed for ^{14}C-activity. The data presented in *Table 1.5* show that considerable proportions of the total ^{14}C-activity which moved out of the application leaf were transported into the xylem stream.

Table 1.5 THE PERCENTAGE DISTRIBUTION OF ^{14}C-ACTIVITY, TRANSPORTED OUT OF THE [14]CO$_2$-APPLICATION LEAF, IN THE BARK, WOOD AND XYLEM EFFLUENT OF STEM SEGMENTS OF WILLOW. (AFTER KING, 1971)

| Experiment | Percentages of total exported ^{14}C-activity | | |
	Bark	*Wood*	*Xylem effluent*
1	29.1	56.4	14.5
2	67.2	16.5	16.3
3	48.7	37.6	13.7
4	44.3	42.1	13.6
5	55.8	20.5	23.7

Excellent evidence for the participation of the symplastic system of the ray cells in radial movement has been provided by Ziegler (1965). He applied ^{14}C-labelled sucrose to the bark of *Robinia* trees and then determined the pathway of centripetal transport by autoradiography. Ziegler demonstrated that radioactivity was confined to the rays and that the velocity of movement was too great to be accounted for by passive diffusion.

All the evidence would seem to point to the conclusion that radial movement in both directions is mediated via a symplastic system. Any apoplastic movement which occurs in intact stems and whole plants is probably negligible.

The tangential transport system

The literature on transport processes contains only small numbers of references to movements of solutes around the stems of plants by tangential transport systems. The few references which have been made indicate that tangential transport only occurs to a slight degree or not at all. O. Biddulph and Cory (1965) studied the movement of ^{14}C-labelled assimilates in *Phaseolus*. Although they observed a considerable radial movement of the ^{14}C-label from phloem to xylem in the apical regions of the plant, very little tangential movement occurred between adjacent vascular bundles. Münch (1937) found an absence of tangential movement between opposite sides of the trunks of trees, while Zimmermann (1960b) failed to demonstrate tangential movement in *Fraxinus americana*.

Nevertheless, it is possible to show that tangential movement can occur under certain conditions, at least in woody plants. Using colonies of the aphid *Tuberolachnus salignus* to provide 'sinks' for movement, Peel (1964) has shown that tangential movement of ^{14}C-labelled sugars can occur in the bark of willow cuttings.

At the present time very little is known about the cellular pathway of this tangential movement. However, certain characteristics of the process strongly indicate a symplastic rather than an apoplastic transport system. Measurements of the velocity of tangential movement in willow have given values as high as 10 cm/h (Peel, 1964). Since these measurements are probably below, rather than above, the true velocity, owing to loss of ^{14}C-labelled sugars from the tangential transport pathway, it is not possible for these values to have been produced as a result of passive diffusion. (For a more detailed analysis of the effect of tracer loss from the transport pathway on the measured velocity of movement, see Chapter 4.) They are, in fact, very similar to those reported by Webb and Gorham (1965a) for radial movement.

To the author's knowledge, there are no reports of tangentially running sieve elements in the bark of willow. It is therefore likely that tangential transport is brought about by a symplastic transfer of solutes between parenchyma cells, accelerated by cytoplasmic streaming. This view is supported by the velocity data referred to above, the values of which are considerably lower than those found for sieve tube movement in willow (Peel and Weatherley, 1962). Also, tangential movement stops at a temperature of $+3°C$ (King, 1971), while sieve tube transport in willow does not cease until the cells are cooled to $-4°C$ (Weatherley and Watson, 1969). The relatively high temperature at which tangential movement is blocked is in accord with a transport system dependent upon cytoplasmic streaming.

2

The range of solutes transported in plants

The determination of mobility

The literature on translocation processes contains many references to the mobility of a great variety of compounds. Before we proceed to enumerate the latter, it is appropriate to examine the methods which have been employed to determine the mobility of compounds in the various transport systems of the plant, for not all the various techniques give unequivocal results.

Gross analytical methods

In the previous chapter (pp. 28–30) some work by Mason and Maskell (1928a) was quoted, in which diurnal changes in the concentration of carbohydrate in the leaves and in the bark and wood of cotton plants were measured. It will be remembered that an excellent statistical correlation was found between the leaf/bark values, particularly when sucrose, rather than total sugar concentrations, was used (*Table 1.3*, p. 30). This led Mason and Maskell to the conclusion that sucrose was the form in which carbohydrates were transported between the leaf source and the bark sink. However, it is possible to criticise this conclusion, for sucrose could merely be a storage compound in the cells at either end of the transport pathway.

In the case of cotton, Mason and Maskell were able to further substantiate the inference concerning the mobility of sucrose, by measurements of the concentration of this compound in the phloem. The presence of high concentrations of sucrose in the transport tissue would seem good evidence for its role in carbohydrate transport. Even so, it could be argued that the sucrose might be present in cells, e.g. parenchyma, other than the sieve elements themselves. Despite the limitations of the gross analytical method, this approach to the determination of mobility proved to be useful before the advent of more modern methods.

Radiotracer methods

In the years since radioactive tracers have become generally available, their use in mobility studies has become widespread. In theory, it is very easy to determine mobilities using labelled compounds; all that has to be done is to apply the radioactive compound to one end of the system, and then after a period of time find out whether activity has moved away from the point of application. In practice, the situation is not so simple, for measurements of radioactivity along the transport pathway have to be combined with analytical techniques to ensure that the label is not moving in a compound other than that which was applied. This may seem an obvious precaution to take, but in fact it was overlooked by some of the earlier investigators. Clearly, however, if the label is confined to the applied compound, then there would be very little doubt that the compound was being transported in an unaltered form.

While there are numerous instances of compounds which do not undergo metabolic change in plants, e.g. many synthetic growth substances and herbicides, when we come to look at the mobility of naturally occurring compounds, the complications introduced by metabolic processes can lead to problems in the interpretation of experimental results. For instance, it may be desired to investigate the form in which carbohydrates are transported in a particular species. To do this the procedure described by Swanson and El-Shishiny (1958) can be employed.

The principles of this method are presented in *Figure 2.1*. Briefly, $^{14}CO_2$ is applied to the leaves of a plant (*Figure 2.1a*); then, after a time has been allowed for assimilation and transport to proceed, the stem is sampled in the direction of movement. Analyses of the distribution of radioactivity in the sugars of the stem samples are made, the data being plotted in the form of distance profiles from the application leaf (*Figure 2.1b*).

The most usual pattern obtained in this type of experiment is that the greatest proportion of the total activity in each stem segment would be present in one sugar (A), and activity in this sugar would extend further down the stem from the point of tracer application. However, activity might also be found in other sugars (B and C), although the distance profiles of these could be shorter than that of sugar A. The ratio, activity in sugar A/activity in sugars B and C, might therefore increase with distance from the application point (Peel, 1966).

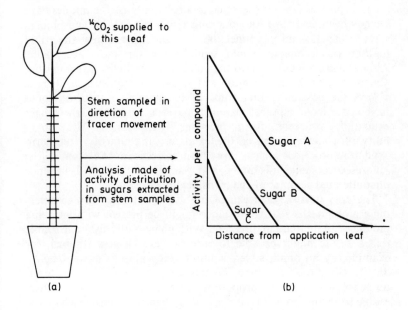

Figure 2.1 Illustration of a technique employing radioactive tracers to determine the mobility of solutes such as sugars. (a) Tracer is supplied to the plant; then, after a period of time, the stem is sampled at increasing distances from the application site. (b) Plots of the distance profiles of the labelled compounds can give information on their relative mobilities

The only firm conclusion which can be drawn from such data is that sugar A is mobile in the transport system, and indeed is the most mobile of the three sugars. Nothing definite can be inferred as to the mobilities of sugars B and C. All that can be said is that B and C are less mobile than A, but in fact they may be completely immobile. Since in such an experiment all the stem tissues have been analysed for activity, sugars B and C might only be present in cells outside the transport conduits, being formed there by metabolism of the mobile sugar A.

Analyses of exudates from transport systems

What appears to be the only method for giving incontrovertible evidence on the mobility of compounds is that in which samples of the transport stream are subjected to direct analysis. Techniques have already been described for obtaining xylem (pp. 22–24) and phloem sap (pp. 32–33) and in connection with the latter system the aphid mouthparts technique would seem to be the most acceptable. Although gross incisions into the phloem undoubtedly give a reasonable picture of the composition of the translocation stream, there is always the possibility that some of the minor constituents could have come from ruptured parenchyma cells. Of course, the longer exudation proceeds from incisions, the less is the possibility of contamination of the exudate. Undoubtedly, long-term exudation processes such as those from the phloem of excised *Yucca* inflorescence stalks (Tammes and van Die, 1966) will produce unadulterated samples of the translocation stream.

The only possible criticism of the use of extracted sap to determine mobilities is that a substance might be present without being mobile in a longitudinal direction. With phloem it is just conceivable that a solute might be able to enter a sieve element, tapped, for example, by an aphid stylet, without being able to move longitudinally along the sieve tube. To the author's knowledge, however, no substance with such properties exists; it would indeed be interesting from the point of view of sieve tube transport mechanisms if one could be found.

In the case of dead xylem vessels, it does not seem possible that a solute, having entered the system, cannot be transported longitudinally in the transportation stream. The only possible barrier to movement might be adsorption on the walls of the vessels. Charles (1953) has suggested that xylem elements may be negatively charged, for he was able to demonstrate rapid petiolar uptake of acid dyes, as compared with the uptake of basic dyes. It seems unlikely that electrostatic adsorption would render a positively charged molecule completely immobile; it would surely just slow down the rate of transport. Transport would thus be analogous to the movement of charged molecules down an ion exchange column, i.e. a series of adsorption and release processes as demonstrated by Bell and Biddulph (1963) for xylem movement of [45]Ca.

Solutes transported in the xylem

Naturally occurring substances

Xylem saps are usually clear solutions having a pH between 5 and 6 and containing a wide range of inorganic constituents, particularly the essential elements (Anderssen, 1929; Bollard, 1953). Bollard (1953) has made a detailed study of changes in the major nutrient composition of xylem sap from apple trees over a period of a year. All the nutrients examined—N, P, K, Mg—showed a low level during dormancy, rising to levels of 160 p.p.m., 30 p.p.m., 175 p.p.m. and 20 p.p.m., respectively, during blossoming time.

Most of the nitrogen taken up by the roots of plants is in the form of nitrate. However, analyses of the nitrogenous fraction of xylem sap has shown that virtually all the nitrogen was in an amide or amino form (Bollard, 1953, 1957a). Bollard (1956, 1957b) has performed a survey of the nitrogen fractions in the xylem sap of a wide range of plants. Nitrate was present in certain species only, and then as a minor constituent.

Two of the major nitrogenous constituents of xylem sap from apple are aspartic acid and asparagine, these compounds making up between 70 and 95 % of the total nitrogen fraction. Other amino acids present (Bollard, 1957a) were alanine, methionine, leucine, serine, threonine and α-amino butyric acid.

Bollard (1960) has listed a number of organic compounds which have been reported to be xylem-mobile, viz.: glutamine, citrulline, allantoic acid, certain alkaloids, phosphoryl choline and cysteine, sugars such as sucrose, organic and keto acids. Enzymes may also be present in xylem saps.

Naturally occurring growth substances seem to be xylem-mobile. Apart from the instances quoted by Bollard (1960) of the occurrence of indole-3-acetic acid (IAA) in xylem saps, there have been a number of recent reports of the presence of the other growth regulators, e.g. gibberellins (Carr, Reid and Skene, 1964) and cytokinins (Kende, 1965). The growth retardant abscisic acid (ABA) has been found in xylem sap from willow by Lenton, Bowen and Saunders (1968) and by Bowen and Hoad (1968). No doubt, the list of naturally occurring substances found in xylem sap will increase considerably as more sensitive detection methods are elaborated.

Synthetic growth regulators, pesticides and antibiotics

During the past 20 years a great deal of work has been done on the patterns of movement in higher plants of a large number of synthetic chemicals and antibiotics produced by micro-organisms, the object being to evaluate these materials as systemic pesticides. A list of some of these chemicals which are known to be xylem-mobile has been given by Bollard (1960, Table 1).

It does not seem necessary here to consider this list in detail, although it is of interest briefly to note the variety of synthetic compounds which can move in the xylem. The range in terms of chemical structure and complexity is very wide, from the relatively simple growth regulators and herbicides, such as 2,4-dichlorophenoxy-acetic acid (2,4-D) and trichloroacetic acid, to complex antibiotics, e.g. streptomycin. It does not appear that there are any absolute criteria by which it is possible to predict the degree of xylem-mobility of a given compound from its chemical configuration; an empirical approach is still necessary in the case of new compounds.

Solutes transported in the phloem

Naturally occurring substances

Crafts and Crisp (1971) have compiled an exhaustive list of the naturally occurring solutes which have been reported to be mobile in the phloem. With the proviso that in certain cases the evidence for phloem-mobility is very slim, in many instances the substances being merely detected in pieces of phloem tissue rather than in the sieve tubes, no better list can be recommended to the reader. However, if we confine ourselves only to those reports in which good evidence is available for mobility, the list of phloem-mobile solutes is still considerable.

Sugars, sugar alcohols, sugar phosphates and nucleotides

One of the most impressive features of sugar movement is the almost universal presence of sucrose as a transportable carbohydrate in phloem of a large number of plant species. In many plants, such as *Salix viminalis*, sucrose may account for over 98% of the total sugars present in sieve tube exudate, being present in concentrations of up to 20% w/v (Peel and Weatherley, 1959). The question must therefore be asked: Why sucrose, rather than some other sugar such as glucose? At the moment we have no answer to this question, although a very ingenious explanation has been put forward by Arnold (1968). He proposed that sucrose acts as a protected derivative of glucose, i.e. a species from which glucose can be derived under certain well-defined conditions, but one which is not subject to extensive enzymic breakdown during transport.

Arnold came to the conclusion that no model based purely on physical parameters would offer more than a marginal advantage for the choice of sucrose over glucose. On the other hand, the protected derivative hypothesis would fit in well with the distribution of enzymes in the plant. The ubiquitous distribution of glucose-catalysing enzymes would render this sugar extremely vulnerable. Sucrose, however, is much less reactive, and therefore this sugar would protect glucose residues from metabolic breakdown until they arrived at sites of growth or storage.

Arnold's conclusions are certainly supported by an examination of the physical properties of solutions of sucrose and glucose, the densities, viscosities and surface tensions of these being so similar that selection on the basis of these parameters is inconceivable. He also concluded that tonicity factors would not be a likely factor in the selection of sucrose. Potentially, the osmotic value of a 20% glucose solution is about twice that of a sucrose solution of the same concentration; therefore it could be argued that sucrose might be inferior to glucose on the basis of its osmotic properties. However, as Arnold points out, the osmotic pressure developed by a solution depends upon the character of the enclosing membrane and the composition of the external solution. In view of our lack of knowledge, it is difficult to make a prediction on the osmotic behaviour of a glucose solution in a sieve tube in place of the natural sucrose solution.

Energy considerations do not, Arnold suggests, help us in this problem. A molecule of glucose can potentially yield 38 molecules of adenosine-5'-triphosphate (ATP). A molecule of sucrose, acted on initially by sucrose phosphorylase, can potentially yield 39 and 38 molecules of ATP, indicating an 'advantage' of only 1/2 an ATP molecule per glucose equivalent in the sucrose solution. Even this small increment would not be realised if the sucrose were hydrolysed by an invertase to free glucose and fructose.

Further consideration was given by Arnold as to why sucrose is the most desirable glucose derivative to employ in translocation. He ruled out many of the glucosides, since the aglycons which would be released on hydrolysis might be toxic. In any event, the glucosides might not readily be metabolised and, hence, might be inefficient energy sources. Arnold reasoned that the disaccharides of glucose would be the most effective in the role of transport sugars, although the reducing disaccharides such as maltose might be less effective, owing to their reactivity.

The presence of oligosaccharides such as raffinose, stachyose and verbascose in phloem exudates from trees, e.g. *Fraxinus americana* (Zimmermann, 1957a), could be explained on the protected derivative hypothesis. These non-reducing sugars have one or more molecules of D-galactose bound to a sucrose residue. Species which have evolved the use of these oligosaccharides have an additional metabolic constraint, for an α-galactopyranosidase would be necessary for complete hydrolysis.

Certainly, Arnold's hypothesis is supported by the evidence that the reactive reducing sugars are very rarely found to be mobile. In a comprehensive survey of the translocation properties of a number of sugars, introduced by a 'flap' technique into the leaves of white ash and lilac, Trip, Nelson and Krotkov (1965) found that the reducing sugars melibiose, galactose, glucose, fructose and pentose were not translocated; the non-reducing sugars sucrose, raffinose, stachyose and verbascose, on the other hand, were readily moved.

However, reducing sugars can clearly be mobile to a limited degree. In exhaustive analyses of sap from excised inflorescence stalks of *Yucca*, Tammes and van Die (1964) found small quantities of glucose and fructose (*Table 2.1*). Certainly, glucose and fructose are mobile within the sieve elements if they can be induced to enter these cells (Peel and Ford, 1968).

There are numerous reports of the phloem-mobility of certain sugar alcohols. Some of these have been found in exudates, e.g. mannitol from white ash (Zimmermann, 1957a); thus there is little doubt concerning their mobility. Trip *et al.* (1965) have shown both mannitol and sorbitol to be mobile in white ash after the application of these labelled compounds to the leaves.

There are only a few instances in the literature of sugar phosphates and nucleotides being detected in exudates, by techniques other than labelling the compounds with ^{32}P. Ziegler (1956) reported the complete absence of hexose phosphates in exudate obtained from a number of tree species. Peel and Weatherley (1959) were unable to detect sugar phosphates in aphid stylet exudate from willow.

On the other hand, Tammes and van Die (1964) (*Table 2.1*) have shown glucose-1-phosphate to be present in *Yucca* exudate. ATP has been found in exudate from incisions (Kluge and Ziegler, 1964) and from aphid stylets (Gardner and Peel, 1969), in concentrations up to 1 mg/ml. Adenosine-5'-diphosphate (ADP) has also been detected in stylet exudate at concentrations approximately one-fifth that of ATP (Gardner and Peel, 1972b).

Sugar phosphates and nucelotides, labelled after supplying [^{32}P] phosphate to phloem, have been reported in extracts of this tissue by Bieleski (1969). The presence of a wide range of ^{32}P-labelled compounds in exudate from *Yucca* stalks and aphid stylets sited in

Salix bark has been demonstrated by Kluge, Becker and Ziegler (1970). These workers have tentatively identified fructose-6-phosphate and fructose-1,6-diphosphate, glucose-6-phosphate, ATP, ADP, uridine diphosphoglucose (UDPG), cytidine-5′-triphosphate (CTP) and phosphoglyceric acid. It seems probable that these compounds are present in low concentrations, and therefore the usual analytical techniques do not reveal them in exudate samples, unless a highly sensitive method such as luciferin–luciferase reaction for ATP can be found. The nucleic acids DNA and RNA have been reported in exudates from *Robinia* (Ziegler and Kluge, 1962).

Table 2.1 SUGARS AND NITROGENOUS COMPOUNDS DETECTED IN EXUDATE FROM THE INFLORESCENCE STALK OF *Yucca flaccida* HAW. (FROM TAMMES AND VAN DIE, 1964, courtesy of the *North Holland Publishing Co.*)

Total dry matter	17.1–19.2%	
Electrical conductivity (20°C)	1.03 mmho/cm	
pH	8.0–8.2	
Sucrose	150–165 mg/ml	
Fructose	2–4 mg/ml	
Glucose	2–4 mg/ml	
Glucose-l-phosphate	*ca.* 1 mg/ml	
Total amino acids (as glutamine)	6.3–10.1 mg/ml	
glutamine		58%
valine		10%
serine/glycine		7%
(iso) leucine		6%
lysine		5%
glutamic acid		4%
-alanine		2%
asparagine		+
aspartic acid		+
Proline	++	
Allantoic acid	+	
Allantoin	+	
Urea	absent	
Total protein	0.5–0.8 mg/ml	

++, trace; +, slight trace.

Nitrogenous compounds

Amino acids form the greater proportion of the nitrogenous fraction of phloem exudates (*Table 2.1*). At least 10 amino acids have been detected in stylet exudate from willow (Mittler, 1958). The most widespread nitrogenous compounds appear to be the amino acids glutamic and aspartic acids and the amides glutamine and asparagine. The concentration of amino acids and amides is not constant throughout the year; seasonal changes occur which have been measured by Mittler in stylet exudate from willow. At the end of dormancy during bud swelling, the total nitrogen concentration was 0.2% w/v, during bud burst it was 0.12%, during leaf maturity 0.03%, rising to 0.13% at leaf senescence. Ziegler (1956) also found a considerable increase in the nitrogenous compounds in exudates from trees during leaf senescence.

Proteins can be present (*Table 2.1*). Concentrations measured by Ziegler (1956) ranged from 0.1 mg/ml in *Robinia* to 1 mg/ml in *Quercus*. This worker considered these to be non-mobile substances, torn out of the sieve tubes during the tapping process. Concentration of protein in cucurbit exudate is much higher than in tree species (Crafts, 1951). We do not yet know whether the protein complement of these exudates is mobile longitudinally, although there seems no reason to doubt the mobility of the soluble proteins. Certainly, it would appear that many of the proteins are enzymic in nature, judging from the large number which have been characterised in exudate from *Robinia* (Kennecke, Ziegler and Fekete, 1971).

Walker and Thaine (1971) have carried out an examination of the proteins in exudate from *Cucurbita* phloem. When gelling of the exudate was prevented by addition of —SH reducing agents, the protein components could be separated into an insoluble or structural fraction which showed an organised fibrillar structure in the electron microscope, and a soluble fraction. The insoluble fraction probably consisted largely of P-protein (Chapter 6) and was probably immobile in the sieve tubes. The soluble fraction contained the gelling protein (termed PS/G protein by Walker and Thaine), which, when allowed to coagulate, produced a gel with no organised fine structure.

Inorganic ions and organic acids

The data presented in *Table 2.2* show that a number of inorganic nutrients are found in phloem exudate from *Yucca*. Potassium is the cation which is usually present in the highest concentration (Peel and Weatherley, 1959). Analyses of anions in exudates, apart from inorganic phosphate and nitrate, have not been carried out to any great extent. It is known (Peel, unpublished data) that both ^{36}Cl and $^{35}SO_4$ are mobile in the sieve tubes of willow; however, cations such as potassium and sodium may have organic anions as their partners. A number of organic acids have been detected in phloem saps, e.g. citric, tartaric and oxalic acids in willow (Peel and Weatherley,

1959) and citric, α-ketoglutaric, maleic, oxalic, succinic and tartaric acids in *Cucurbita* (Kating and Eschrich, 1964).

Table 2.2 IONS DETECTED IN EXUDATE FROM THE INFLORESCENCE STALKS OF *Yucca flaccida* HAW. (FROM TAMMES AND VAN DIE, 1964, courtesy of *North Holland Publishing Co.*)

	mg/ml
Total phosphorus	0.310
Inorganic phosphorus	0.105
Nitrate	absent
Potassium	1.680
Magnesium	0.051
Calcium	0.014
Sodium	0.004 1
Zinc	0.002 1
Iron	0.004 1
Manganese	0.000 5
Copper	0.000 4
Molybdenum	0.000 01

Other substances

Steroids have been implicated as translocates in *Phaseolus* (O. Biddulph and Cory, 1965). Naturally occurring growth substances have been detected in phloem sap: IAA from *Fagus sylvatica* (Huber, Schmidt and Jahnel, 1937), gibberellins from *Robinia* (Kluge, Reinhard and Ziegler, 1964) and from *Taraxacum* (Hoad and Bowen, 1968), and abscisic acid from willow (Hoad, 1967). Vitamins such as nicotinic, pantothenic and folic acids have been detected in exudates from *Robinia* (Ziegler and Ziegler, 1962).

The list of naturally occurring substances in phloem exudates grows larger year by year. No doubt, as with the xylem saps, the list of compounds will continue to increase as more refined techniques of analysis are elaborated. There appears to be no reason why any water-soluble compound should not find its way into the sieve tubes, albeit in some instances in extremely low concentrations.

Synthetic compounds

Much of the work on the movement of synthetic compounds in the phloem has been performed by applying the compound in a labelled form to either the roots or the leaves. After a certain time had been allowed for uptake and transport, the plant was harvested and freeze-dried, and autoradiographs were prepared of the whole plant to determine the distribution of the label (Crafts and Crisp, 1971). While this work has produced evidence for the phloem-mobility of a considerable number of synthetic compounds such as 2,4-D, maleic hydrazide and pichloram, very little work has been published which demonstrates that these compounds can move into the sieve elements.

In a series of experiments performed on willow, in which sieve tube exudate was obtained via the stylets of aphids, Field and Peel (1971b, 1972) reported the results of their experiments in which the sieve tube mobilities of a range of synthetic compounds were studied. All the compounds used—1-naphthaleneacetic acid (NAA), 2,4-D, 4-chloro-2-methylphenoxyacetic acid (MCPA), 2,4,5-trichlorophenoxyacetic acid (2,4,5-T), trichloracetic acid (TCA), maleic hydrazide and paraquat—were found to enter the sieve elements and to move longitudinally through the sieve tubes.

The point to stress at the moment is that, clearly, a number of compounds of widely different chemical structures can enter the sieve elements and are able to move through these cells. We do not yet know how these synthetic compounds are able to enter the sieve tubes, but we shall be returning briefly to this topic in Chapter 3.

Solutes transported in radial and tangential transport systems

Unlike the xylem and phloem, it is not possible with radial and tangential transport systems directly to sample the solute stream, since movement takes place here in relatively non-specialised cells. It is not surprising, therefore, that our knowledge of the mobilities of solutes in these pathways is very limited, such information as we possess having been obtained from chemical analyses of the tissues, or from radiotracer experiments using the technique illustrated in *Figure 2.1.*

To avoid spending a great deal of time on this subject, it seems possible to make the general statement, based on the scant information available, that solutes mobile in the phloem are also mobile on the radial and tangential pathways, e.g. of the sugars, sucrose seems to be mobile. Zimmermann (1958) has produced evidence for the radial movement of sucrose from the sieve tubes of *Fraxinus*. Webb and Gorham (1965a) concluded that sucrose could move from the sieve tubes of cucurbit phloem into surrounding tissues, while Peel (1966) showed that sucrose was more mobile than either glucose or fructose in the tangential transport system of willow stems.

Ions can move freely out of the phloem ($[^{32}P]$phosphate and $[^{35}S]$ sulphate, S. F. Biddulph, 1956; ^{86}Rb, ^{22}Na, Peel, 1967), and between the xylem and phloem (see Chapter 1). Radial movement of growth substances has been reported by Bowen and Wareing (1969) using ^{14}C-labelled kinetin and gibberellic acid. Radial movement of IAA can occur readily in willow stems (Field and Peel, 1971a); also, tangential movement has been reported (Lepp and Peel, 1971c; Leach and Wareing, 1967).

Similarly, synthetic compounds which are mobile in the phloem can move in a radial direction. Eliasson (1965) and Sunderam (1965) have demonstrated radial movement of certain of the chlorophenoxy herbicides in trees. Field and Peel (1971a) have given data showing that NAA, the chlorophenoxy herbicides, TCA, maleic hydrazide and paraquat can move both centrifugally and centripetally in willow stems.

The similarities which exist in the types of compounds transported, between the phloem on one hand and the radial and tangential systems on the other, might point to certain basic features common to the three transport processes. While this may be so, it would be presumptuous to suggest that these similarities indicate that the three systems possess the same mechanism of transport.

3

The control of solute loading and unloading at sources and sinks

In the present chapter we must attempt to answer the question as to what mechanisms control the rate at which solutes enter or leave the sieve elements. A second problem which also must be considered is that posed by some of the data presented in Chapter 2: How are some solutes able to enter the sieve elements while others are partially or wholly excluded? As will become apparent, our knowledge of this aspect of phloem physiology is considerably more meagre than that concerning the over-all control of transport, which we discuss in Chapter 5. It must be clear, however, that an understanding of the control of solute loading and unloading is essential if we are ever going to unravel the complex problems posed by phloem. To give a simple illustration, there are, to the author's knowledge, no solutes which are able to enter the sieve elements which cannot subsequently be transported. On the other hand, there are a number of instances of solutes (e.g. glucose and fructose—Peel and Ford, 1968) which cannot normally enter the sieve elements but which can be induced to enter by experimental manipulation. In these cases the 'abnormal' solutes can apparently move quite freely through the sieve tubes, a fact which is of obvious importance to any consideration of the mechanism of longitudinal transport.

General considerations and the concept of solute 'potential'

There seems little doubt that active, energy-requiring processes are responsible for the movement of solutes into the sieve elements.

Therefore it is to be expected that movement will exhibit the various characteristic features associated with active transport mechanisms; specificity and competition may occur between chemically related solutes, and movement could occur against concentration or electro-chemical potential gradients. The rate of loading, therefore, need not necessarily be related to concentration gradients between the source tissues and the sieve tube, except probably in so far as the rate of an enzymic process is governed by the concentration of the substrate.

Some evidence for the view that movement of sugars into the phloem is an active process, rather than one mediated by diffusion, was obtained by Phillis and Mason (1933) from their work on sugar gradients in cotton leaves. Analysis of sucrose, hexose and poly-glucoside concentrations in sap obtained from the mesophyll and veins revealed that only the polyglucosides possessed a positive gradient (*Table 3.1*).

Table 3.1 CONCENTRATIONS (g/100 cm³ OF SAP) IN VARIOUS PARTS OF THE LEAF OF COTTON. (FROM PHILLIS AND MASON, 1933, courtesy of *The Clarendon Press*)

| | Lamina | | Petiole | |
	Mesophyll	*Vein*	*Bark*	*Wood*
Sucrose	0.61	1.04	1.06	1.13
Glucose	0.41	1.15	0.42	0.71
Fructose	0.23	0.20	0.09	0.29
Polyglucoside	0.29	0.02	0.05	0.01

Phillis and Mason considered it unlikely, in spite of the positive gradient for polyglucoside, that sugars were transported from the mesophyll cells into the sieve tubes in this form, since its concen-tration showed no response to ringing and it did not disappear from the leaves after a long period of darkness.

Kursanov (1963) has reported on a wide range of translocation experiments performed by a group of Russian workers. Of particular interest in the present context are the results of certain work on the effect of ATP on movement of ^{14}C-labelled assimilates out of leaves of sugar beet. Kursanov and Brovchenko (quoted in Kursanov, 1963) allowed the leaves to assimilate $^{14}CO_2$ for 5 min, then infiltrated them with either water or a $6 \times 10^{-3}M$ solution of ATP. Treatment with ATP increased the amount of labelled organic (particularly hexose) phosphates in the leaves as compared with the water con-trols. More important, ATP enhanced the export of assimilates by up to seven times that of the controls in young beet plants.

These facts could readily be explained on the basis of an enhance-ment of the loading process by ATP. Certainly, since neither ADP nor adenosine-5'-monophosphate (AMP) stimulated transport, it must be presumed that it was the third phosphate group of ATP which was supplying energy to the loading mechanism.

Further evidence that movement into phloem tissues involves active processes has been produced by Bieleski. Working initially upon slices of sugar cane pith which contained vascular tissues, Bieleski (1962) demonstrated that the vascular tissues accumulated sugars from the external milieu at a rate 5 to 20 times greater than that of the pith parenchyma, autoradiographic evidence suggesting the phloem tissues as being responsible for most of this enhanced accumulation.

In a later paper Bieleski (1966) reported the results of work on isolated vascular bundles from a number of species in which he studied the uptake of phosphate, sulphate and sucrose. The former ion was accumulated principally as inorganic phosphate, the sucrose mainly as sucrose. Rates of accumulation were considerably greater than into parenchymatous tissue.

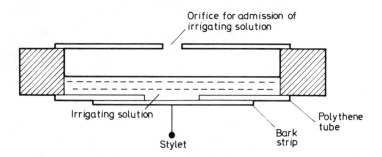

Figure 3.1 Willow bark strip system for studying solute movement into sieve elements. Sieve tube sap is collected via severed aphid stylets

Despite the fact that sieve tube loading is patently an active process, evidence is available which shows that the rate of loading can be affected by the magnitude of the concentration gradient between the source cells and the sieve elements. It has been shown by Weatherley *et al.* (1959) that the application of osmotica to stem tissues, e.g. bark strips of willow on which aphid stylets were exuding (*Figure 3.1*), led to a fall in the rate of sucrose exudation accompanied by an increase in the concentration of the exudate. It was suggested that sucrose secretion into the sieve tube was initiated by the dilution of the sieve tube contents consequent upon stylet puncture, sucrose transfer rates then being a function of the sucrose concentration in the sieve tube. *Figure 3.2* shows the relationship between the sucrose flux and sucrose concentration from a single stylet sited on a bark strip which was bathed with successively higher concentrations of mannitol (Peel, 1972b).

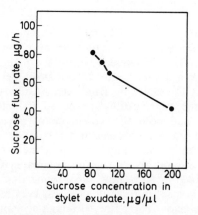

Figure 3.2 Relationship between sucrose exudation rate and sucrose concentration in exudate from aphid stylets during irrigation of cambial surface of a bark strip with mannitol solutions of increasing osmotic potential. (From Peel, 1972b)

More precisely, Weatherley and his colleagues envisaged that movement both into and out of the sieve tube is governed by differences in sucrose 'potential' This would be related to, though not necessarily equated with, concentration, since movement is active and therefore could take place against a concentration gradient. Regions of sucrose utilisation would have a low sucrose 'potential' and therefore would act as sinks. Regions of synthesis or mobilisation of reserves would have a high sucrose potential and therefore would constitute sources, secreting sugar into the sieve tubes.

Further evidence that the 'potential' of a solute may not be related to its concentration was obtained by Peel (1963), using potassium and sodium. Perfusion of the xylem of stem segments of willow with solutions of these ions frequently produced an increase in the concentration of the ion in sieve tube exudate. On the other hand, instances were found where an increase in the concentration of an ion in the segment tissue did not produce a concomitant increase in concentration in the sieve tube.

Hartt (1963) has shown that an accumulation of sucrose in sugar cane blades, produced by a reduction in translocation, can reduce the rate of photosynthesis. Here the sucrose 'potential' in the sieve tubes would be high; therefore the movement from the mesophyll would be low, which would lead to an accumulation of photosynthetic products in the latter cells.

Selectivity of the sieve tube

As we have seen in Chapter 2, many different substances may be found in sieve tube exudates, but there are also a number of chemical species which, although present within the plant, enter the sieve elements only to a limited degree or not at all. Several examples of the specificity of the sieve tube have been cited by Kursanov (1963). In a study of the translocation of labelled assimilates by leaves of rhubarb which had been allowed to photosynthesise in $^{14}CO_2$, it was shown that organic acids tended to accumulate on the boundary of the conducting tissues, only small quantities of malic and citric acids actually entering the translocation stream. Kursanov and his colleagues (quoted in Kursanov, 1963) found that the selectivity of amino acid movement into the phloem was even more marked than that of organic acids. Threonine, although accounting for 25–35% of the total amino acid in the veins of rhubarb leaves, only carried 3–4% of the activity in the mesophyll tissue. Serine and alanine were also found to enter the conducting tissues with ease, although aspartic acid and proline were appreciably less mobile.

Many other instances may be quoted of the selective nature of sieve tube uptake processes, particularly with regard to the uptake of organic solutes. In a study of sugar movement between the source cells and the sieve tubes of bark strips of willow, Peel and Ford (1968) presented data (*Table 3.2*) illustrating the highly selective nature of sugar movement into the sieve elements; sieve tube sap obtained via severed aphid stylets 24 h after irrigation of bark strips with labelled sugars contained virtually all the activity in sucrose, although extracts of the whole bark revealed considerable amounts of activity in hexose and organic phosphates.

Selectivity in uptake is shown by ions as well as organic solutes. Judging from the results of a number of studies on the uptake and transport of foliar applied isotopes (quoted by Zimmermann, 1960a), it seems clear that potassium and sodium are the most mobile, while the alkaline earth cations such as calcium are the least mobile. Of course, many factors must inevitably influence the mobility of substances applied to the leaf, and it is not certain from such studies whether the relative immobility of calcium is due to an inability to penetrate rapidly into the sieve elements or whether some other factor might be operating. If calcium is applied to plants, then much of it becomes rapidly immobilised in a water-insoluble form. Wiersum, Vonk and Tammes (1971) have suggested that this is the prime reason for the relative immobility of this ion.

Table 3.2 THE DISTRIBUTION OF ^{14}C-ACTIVITY IN APHID STYLET EXUDATE AND BARK EXTRACTS 24 h AFTER APPLICATION OF LABELLED SUGARS TO A BARK STRIP OF WILLOW. (FROM PEEL AND FORD, 1968)

Sample		Sucrose experiment	Glucose experiment
Stylet exudate	organic phosphates	nil	0.22
	sucrose	2.28	5.42
	other sugars	nil	nil
Bark extract	organic phosphates	2.04	3.20
	sucrose	3.00	0.92
	glucose	5.94	11.44
	fructose	4.92	7.80

Activities given in nCi.

However, there is some evidence (Peel, 1972b) that calcium cannot enter the sieve tube, at any rate in willow, as readily as can the phloem-mobile cation caesium. Solutions of $^{45}CaCl_2$ and $^{137}CsCl$, both at the same specific activity, were introduced into the xylem of stem segments. Sieve tube exudate was collected as honeydew from whole individuals of *Tuberolachnus*; thus the mass of each ion moving into the sieve elements in a given time could be measured. At the end of the experiment the total water-soluble mass of each ion in the bark was determined; thus the relative mobility of each ion could be measured by the ratio, mass in honeydew/mass of water-soluble ion in bark. The data presented in *Table 3.3* demonstrate that calcium enters the sieve elements less readily than does caesium. Thus it would appear that, in willow at all events, the relative immobility of calcium is not only caused by the conversion of this substance into a water-insoluble form.

Table 3.3 THE RELATIVE MOBILITIES OF ^{45}Ca AND ^{137}Cs BETWEEN THE BARK CELLS AND SIEVE ELEMENTS OF WILLOW. (FROM PEEL, 1972b)

Experiment	Relative mobility	
	^{45}Ca	^{137}Cs
1	0.002	0.025
2	0.001	0.013
3	0.005	0.021
4	0.008	0.017
5	0.001	0.026

We do not yet understand the mechanisms whereby the sieve tubes are able to discriminate between different solutes. However, from the results of experiments on sugar transport from the solution irrigating the cambial surface, into the sieve elements of bark strips of willow, Peel and Ford (1968) suggested that the tonoplast of the source cells might be the site of the selection process. On the basis of observations on phloem structure, Peel and Ford concluded tha t two pathways for sugar movement could exist in bark strips: a 'direct' one via the companion cells, in which the sugars would not have to traverse a tonoplast, and an 'indirect' one via the storage (source) cells of the bark, in which a tonoplast might have to be crossed.

As a corollary to this work, Peel and Ford concluded that movement on the 'indirect' pathway was mediated through the plasmodesmata. Webb and Gorham (1965a) had reached the same conclusion with regard to movement out of the phloem. However, evidence is available from some studies which indicates that movement of sugars may occur through the free space, i.e. cell walls and intercellular spaces. Cormack and Lemay (1963) have given results which are consistent with free space movement of labelled sugars in white mustard roots. Hawker (1965) found that the concentration of sucrose in the volume external to the cell vacuoles of sugar cane stems was very nearly as great as that in the vacuoles. This 'external' sucrose was situated mainly in the aqueous phase of the cell walls and intercellular spaces. Hawker concluded that the high free space concentration strongly favoured the view that intercellular transport occurs mainly via the cell walls rather than through the plasmodesmata. It could well be that the pathways for solute transport into the sieve tubes differ between species.

The mechanism of sugar transport into sieve tubes

A number of schemes have been proposed to explain sucrose transport, and it is necessary to examine the relative merits of these on the basis of the experimental information which is available.

The first problem to be considered is whether sucrose is hydrolysed during its transport between the sieve tubes and the source (or sink) cells. Some evidence is available from work with sugar cane (Sacher, Hatch and Glasziou, 1963) indicating that sucrose is acted upon by invertases during its transport out of the conducting tissues. In immature tissues an approximately linear relationship was found between the rate of growth and the invertase content of the tissue (Hatch and Glasziou, 1963). This invertase had an optimum activity between pH 5.0 and 5.5, and was apparently located in two separate compartments, the outer (free) space and the storage compartment (*Figure 3.3*). Sacher and his colleagues suggested that the function of the enzyme in the outer space was to control the rate at which sucrose was moved from the conducting tissue to the young, growing cells, while that in the storage compartment was concerned with the remobilisation of sucrose after storage. In mature tissue with a high capacity for sucrose storage, the acid invertase was replaced by a neutral invertase with a pH optimum of 7.

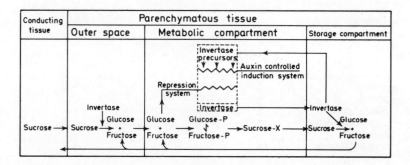

Figure 3.3 Schematic representation of the sugar accumulation cycle in sugar cane storage tissue. Glucose-P, fructose-P are phosphorylated hexoses and sucrose-X is a sucrose derivative, which it is presumed provides energy by its cleavage for the accumulation of sucrose against a concentration gradient into the storage compartment. (From Sacher, Hatch and Glasziou, 1963, courtesy of the American Society of Plant Physiologists)

Further evidence for these conclusions concerning the role of invertases in the partitioning of assimilates may be adduced from the observation that enzyme levels can be affected by external conditions. Working with sugar cane tissue slices, Hatch and Glasziou (1963) demonstrated a rapid decline in storage compartment invertase when either glucose or fructose was applied. The level of this invertase could be raised by application of auxin (*Figure 3.3*).

Ford and Peel (1967b) investigated uptake of labelled sugars into the sieve elements of bark strips of willow. They found that uniformly labelled [14C]sucrose was rapidly hydrolysed in the solution bathing the cambial surface of the strip, presumably by a free space invertase. The labelled sucrose which appeared in the stylet exudate had clearly been inverted and then reconstituted, since its glucose/fructose activity ratio differed from that of the sucrose which was applied to the strip. However, Ford and Peel were unable to show that inversion was a prerequisite for uptake.

Hatch and Glasziou (1964) applied fructosyl-U[14C]sucrose (sucrose labelled only in the fructose moiety) to the cut end of the midrib of sugar cane leaves for 8 h. After this time, the radioactive compounds in the midrib, sheath and other tissues were examined. The results shown in *Table 3.4* demonstrate that there was little randomisation of the 14C-label in the hexose moieties of sucrose, except in the case of the internodal tissue.

Table 3.4 TRANSLOCATION OF FRUCTOSYL-U[14C]SUCROSE FROM LEAF TO SHEATH AND STORAGE TISSUE OF STEM. (FROM HATCH AND GLASZIOU, 1964)

Tissue	% total radioactivity			Ratio of 14C in glucose to fructose moieties of sucrose
	Other compounds	Sucrose	Glucose + fructose	
Midrib	2	96	2	0.05
Sheath	3	95	2	0.03
Nodal tissue	9	90	1	0.06
Internodal tissue	6	92	2	0.09
Juice from internodal tissue	14	76	10	0.12
Fibre from internodal tissue	1	97	2	0.03

Hatch and Glasziou interpreted these results to mean that the sucrose molecule is translocated intact through the leaf, sheath and stem of sugar cane. The randomisation of the label in the hexose moieties of sucrose from parenchyma tissues, and the relatively high proportion of radioactivity in glucose, fructose and other compounds, were, these authors conclude, consistent with the translocated sucrose being hydrolysed upon leaving the transport conduits.

Evidence for the inversion of sucrose, during either its passage out of, or movement into, the sieve tubes, appears quite convincing. It must be remembered, however, that much of the evidence for inversion as an essential step in movement comes from work almost exclusively confined to sugar cane. The situation could be quite different in other species.

In certain trees oligosaccharides are present, often making up a large proportion of the total sugars. Zimmermann (1958) has demonstrated changes in the concentration of these sugars in phloem exudate obtained from *Fraxinus americana* (L.) after defoliation. Stachyose decreased continuously to almost complete disappearance; sucrose increased after an initial drop, presumably at the expense of stachyose. Raffinose, an intermediate between these two sugars, always stayed at a very low level (*Figure 3.4*). In a later paper Zimmermann (1961) suggested that an α-D-galactosidase, situated in the side wall cytoplasm of the sieve tubes, was responsible for the breakdown of stachyose to sucrose. Since galactose did not appear in the exudate, sugar must be transferred directly out of the sieve tube lumen as galactose units.

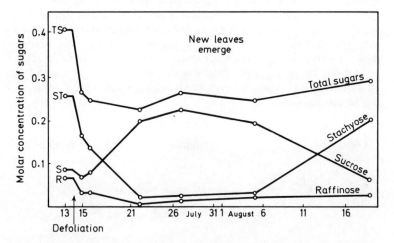

Figure 3.4 The effect of defoliation on the molar sieve tube sugar concentrations (vertical axis) at 5 m height, the approximate middle of the entire phloem length. (From Zimmermann, 1958, courtesy of the American Society of Plant Physiologists)

It may have been noted in *Figure 3.3* that a compound 'sucrose-X' was implicated by Sacher *et al.* (1963). There are a number of reports in the literature which suggest that 'sucrose-X' may be sucrose phosphate and that this compound is intimately concerned in the movement of sucrose between cells.

Kursanov (1963) quotes some work by Barrier and Loomis (1957), in which the latter workers showed that sucrose enhanced the uptake of ^{32}P-labelled inorganic phosphate by soya leaves. These findings, according to Kursanov, could well indicate a relationship between the uptake of sucrose and of phosphate, i.e. the two could be moving together as sucrose phosphate.

The main difficulty in envisaging a role for sucrose phosphate in sugar transport lies in the fact that it is not possible to detect this compound in many leaves. However, as Kursanov points out, this may be due to its rapid desphosphorylation in the phloem. As previously mentioned, Kursanov and his co-workers performed experiments in which they enriched sugar beet leaves with ATP, but this did not lead to the detection of sucrose phosphate. However, ATP accelerated the uptake of monosaccharides by the phloem more than that of sucrose. Kursanov concludes that these data seem to favour the view that transport of sucrose from the mesophyll into the phloem is preceded by its inversion and the phosphorylation of its constituent hexoses, the latter entering the sieve elements, either directly or by means of special carriers, there being rapidly resynthesised to sucrose. Kursanov has given a scheme (*Figure 3.5*) summarising his views of the possible biochemical transformations involved. According to this, sucrose phosphate might only participate in transport at the chloroplast membrane and in sucrose resynthesis in the phloem; therefore the mass of this compound in an exporting leaf would be very small, which would lead to difficulties in detection.

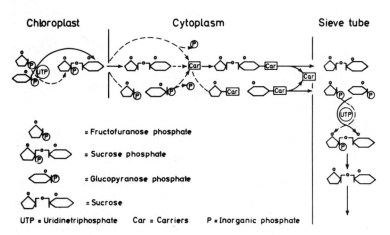

Figure 3.5 Transport scheme for sugars from chloroplast into sieve tube. (From Kursanov, 1963, courtesy of Academic Press)

Certainly, sucrose phosphate has been found in plant tissues. Hatch (1964) reported some experiments on the role of this compound in sugar uptake by sugar cane storage tissue. When U[14C] glucose was supplied to tissue slices, a compound with the properties of sucrose phosphate became labelled; i.e. when the compound was treated with phosphatase and invertase, labelled glucose and fructose were formed. Hatch also investigated the enzymes involved in the synthesis of sucrose phosphate. He was unable to detect the presence of a sucrose kinase, although uridine diphosphate glucose–fructose

6-phosphate glycosyl transferase was found, which can catalyse sucrose phosphate formation from UDP-glucose and fructose 6-phosphate. He suggested sucrose phosphate as the compound designated 'sucrose-X' in *Figure 3.3*.

Some experiments by Gardner and Peel (1971) provided no evidence for the participation of sucrose phosphate in sugar uptake by sieve elements of willow. After application of ^{14}C-labelled sucrose to a bark strip, activity appeared in aphid stylet exudate, not only in sucrose and to a lesser degree in hexoses but also in organic phosphates. Chromatographic analysis of the latter compounds showed that UDP-glucose and UDP were labelled. No evidence was found for the presence of labelled sucrose phosphate in the stylet exudate, although the possibility that it might participate in sugar transport could not be ruled out completely, for Lester and Evert (1965), Kursanov (1963) and Ulbrich (1969, quoted in Kennecke *et al.*, 1971) have all detected phosphatases in phloem; thus any sucrose phosphate might rapidly be converted to sucrose as it entered the sieve tube.

Gardner and Peel concluded that sucrose probably enters a metabolic compartment (companion cells?) by traversing the plasma membrane as hexose. Resynthesis of sucrose in this metabolic compartment could be accomplished via a UDP, UDP-glucose system, followed by movement into the sieve element lumen.

The movement of synthetic compounds into sieve tubes

It has been mentioned in Chapter 2 that a wide range of synthetic compounds can enter the phloem and be transported longitudinally through this tissue. However, we have very little information on how the movement of these substances is controlled. It would appear that the only way of investigating the control of movement would lie in the use of techniques in which sieve tube exudate can be examined. Thus, by applying a known mass of a substance to the bark, it would be possible to measure the rate at which it entered the sieve elements.

An attempt to measure the relative mobilities of certain growth regulators and herbicides has been made by Field and Peel (1971b). They applied labelled synthetic compounds to bark strips of willow for 16 h and then removed the excess by washing the cambial surface of the strip. After establishing an exuding stylet on the strip, they measured the specific activity of the exudate, after which the total activity in the bark was assayed. The relative mobilities of certain of the compounds are given in *Table 3.5*, maleic hydrazide and MCPA having the highest mobilities, and 2,4,5-T and paraquat the least.

Table 3.5 RELATIVE MOBILITIES OF CERTAIN GROWTH REGULATORS AND HERBICIDES IN BARK STRIPS OF WILLOW. (FROM FIELD AND PEEL, 1971b, courtesy *Blackwell*)

Compound	*Relative mobility (specific activity of stylet exudate/total activity in bark)*
Maleic hydrazide (MH)	0.022
Methyl chlorophenoxyacetic acid (MCPA)	0.022
Trichloroacetic acid (TCA)	0.017
2,4-Dichlorophenoxyacetic acid (2,4-D)	0.012
2,4,5-Trichlorophenoxyacetic acid (2,4,5-T)	0.005
Paraquat	0.002

One of the most interesting points arising from these experiments is the fact that all the substances employed, even paraquat, which is generally thought of as being mobile only in the xylem (Funderburk and Lawrence, 1964), entered the sieve elements. Indeed, the only substances which Field and Peel found to be immobile were two metabolites of naphthaleneacetic acid. Another most unexpected result was that all the compounds were loaded into the sieve elements; when a solution of a compound was introduced on to the cambial surface of a bark strip bearing an exuding stylet, the specific activity of the exudate exceeded that of the irrigating solution within 4–6 h from application.

This latter observation precludes the possibility that movement could have been merely diffusional. Therefore enzyme systems must be involved, but the question is: What enzyme systems? It seems unlikely that new enzymes would be formed to load the synthetic compounds, in view of the short application times in the experiments of Field and Peel. It therefore seems reasonable to propose that existing enzymes are utilised, which normally are involved in the transport of naturally occurring solutes. If this proposition is correct, it would be expected that competition between a synthetic compound and its natural analogue might occur. Some evidence that competition occurs, at least in the case of maleic hydrazide, has now been produced by Coupland and Peel (1972), who demonstrated that this compound will inhibit uracil uptake by sieve tubes.

So far, only a few brief references have been made to the control of solute unloading from sieve tubes. The reason for this is simple: we know even less concerning this process than we do about loading.

Certainly, it would appear that the concept of 'potential' is of use in relation to the unloading process. Hoad and Peel (1965a) have demonstrated an effect of changing the concentration of sugars and phosphates in willow plants on the rate of lateral loss of these solutes from the sieve tubes to the xylem. With 'low' sugar or phosphate plants, the rate of loss was high, the situation being reversed in 'high' solute plants.

There is some evidence that unloading may differ from loading in the energy relations of the process. We have already referred to experiments which demonstrate loading to be an energy-requiring process, but it appears that unloading may not be so energy-dependent. For instance, King (1971) has demonstrated that movement of ^{32}P-labelled phosphates into willow sieve elements ceases at 2°C. On the other hand, Ford and Peel (1967a) showed that this temperature did not inhibit the loss of ^{14}C-labelled sugars from the sieve tubes of this species.

Clearly, a large amount of work needs to be carried out if we are to gain a clear understanding of the fundamental physiological and biochemical processes underlying movement into and out of sieve elements. However, it is hoped that this brief review has given the reader a clear picture of the importance of these processes, not only to a clarification of phloem physiology but also to the functioning of the whole plant.

4

The measurement and concepts of velocity and mass transfer

There are many instances in the literature on transport of the measurement of the 'rate' of translocation. The term 'rate', however, is abused, for what most workers mean in this context is velocity. It seems quite natural to picture transport taking place through the sieve tubes as a solution moving with a given linear velocity having the unit of cm/h. Of course, it is not supposed that the measured velocities give anything but an average velocity of the moving molecules. There are no moving fluid cylinders in nature; velocities are always graded, in the case of ideal capillaries along the surface of a paraboloid.

Canny (1960b) in an excellent review has questioned the wisdom of using velocity as a useful parameter of translocation systems. He believes that it is not necessary to view the transport process in the sieve tubes as consisting of the movement of a solution. However, the moving solution, if it exists, must presumably possess some average velocity which is related to the volume transfer:

$$\text{volume transfer} = \text{area} \times \text{velocity} \qquad (4.1)$$
$$(\text{cm}^3/\text{h}) \qquad (\text{cm}^2) \qquad (\text{cm}/\text{h})$$

Now, this hypothetical solution must possess a certain concentration; therefore we could express the 'rate' of movement, not in units of volume transfer but in terms of mass transfer, thus:

$$\text{mass transfer} = \text{volume transfer} \times \text{concentration} \qquad (4.2)$$
$$(\text{g}/\text{h}) \qquad (\text{cm}^3/\text{h}) \qquad (\text{g}/\text{cm}^3)$$
$$= \text{area} \times \text{velocity} \times \text{concentration}$$
$$(\text{cm}^2) \qquad (\text{cm}/\text{h}) \qquad (\text{g}/\text{cm}^3)$$

or

$$\text{specific mass transfer} = \text{velocity} \times \text{concentration} \qquad (4.3)$$
$$(\text{g cm}^{-2} \text{ h}^{-1}) \qquad (\text{cm}/\text{h}) \qquad (\text{g}/\text{cm}^3)$$

It is Canny's contention that the difficulties involved in the measurement of velocities are very great, and moreover, as we have already noted, velocity implies that transport takes place in a moving solution, with which concept some phloem workers disagree. If, however, we use specific mass transfer as a measure of the magnitude of the transport process, then this parameter of the system could be employed by all workers, irrespective of their allegiance to any particular hypothesis of the mechanism of movement. It is not necessary to employ equation 4.3 to derive the specific mass transfer values of a system from measurements which may be attempted of the velocity and concentration of a hypothetical solution; in certain systems determinations of specific mass transfer may be made directly.

The measurement of velocity

Chemical methods

Studies on sugar concentrations in the phloem have revealed a diurnal cycle of change in extracts of bark or in exudates collected from incisions into the phloem (Mason and Maskell, 1928a; Huber *et al.*, 1937). Sugar concentration was shown to be lowest in the early morning, rising during the day in response to the progress of photosynthesis to a maximum in the late afternoon, then falling again during the evening and night. This 'concentration wave' phenomenon has become generally accepted as an indication of phloem transport. Huber *et al.* (1937) sampled the exudate from cut phloem at different heights on the trunk of a red oak, and showed that a wave form of sugar concentration seemed to appear at progressively lower heights during the course of the evening and night. From the data presented in Figure 4 of their paper, Huber and his colleagues calculated that the minimum sugar concentration was moving down the phloem of the tree at a velocity of 3.6 m/h.

Such a velocity is extremely high, and the validity of these measurements has been questioned on theoretical grounds by Canny, Nairn and Harvey (1968). Clearly, with such a velocity the mass transfer would be enormous, unless a solution was moving with a very low concentration (equation 4.2). Canny and his co-workers argued that the thickness of active phloem in the bark of a red oak with a 12 m trunk would not be less then 3 mm, and the radius of the trunk not much less than 30 cm. The phloem area would then be approximately 57 cm^2; and if it is assumed that only one-fifth of this is sieve tube area, then the area undertaking transport would be

57/5 cm². Substituting the data of Huber *et al.* for velocity and concentration of the moving solution in equation 4.2, the mass transfer is

$$\frac{57}{5} \times 360 \text{ (velocity)} \times 0.18 \text{ (concentration)}$$

$$= 740 \text{ g/hr} \text{ or } 17\,800 \text{ g/day}$$

If the net assimilating capacity of the leaves is about 0.7 g $m^{-2}h^{-1}$ and the leaves photosynthesised for 12 h per day, then the leaf area required to produce this flow rate would be

$$\frac{17\,800 \times 2}{0.7} = 51\,000 \text{ m}^2$$

Since the leaf area index is unlikely to be more than 3, the projected area of this tree on the ground would be about 17 000 m² or over 4 acres! As Canny *et al.* point out, this figure is so patently ridiculous that any errors in their assumption must be rendered negligible; the value for velocity obtained by Huber *et al.* is therefore an impossibility.

In fact, as mentioned by Zimmermann (1969), Huber and his collaborators were well aware of at least two objections to their use of the 'concentration wave method' for measuring velocities. It could be that the concentration wave does not show phloem transport, but may be merely a reflection of a descending tension wave within the xylem. Also, even if the wave indicated phloem transport, it could be distorted by the effect of xylem tensions upon the sieve tube exudate. Certainly, changes in water potential in the xylem can markedly affect the concentration of phloem exudates obtained via aphid mouthparts (Weatherley *et al.*, 1959; Peel and Weatherley, 1962). However, Huber and his co-workers dismissed the idea that xylem tensions could affect the concentration wave to such a degree as to render their measurements useless.

Zimmermann has made two ingenious approaches to the problem of measuring velocities in trees using chemical analyses on exuded sap. In 1958 he extracted a value for the velocity of movement in American ash from data on the changes in sugar concentration following defoliation. With leaves present, a gradient of sugar concentration down the trunk was found of 17% (of the value at the top of the tree) in 8 m. After defoliation there was a rapid fall in the sugar concentration at an intermediate height of about 1.3% per hour of the initial value. Zimmermann reasoned that the initial gradient was the result of loss of sugar from the sieve tubes, the magnitude of this process being revealed when the supply of sugar from the leaves was removed by defoliation. He further argued that if this rate of removal was normal, then the decrease of

17% would have required 17/1.3 or 13 h to travel 8 m. Therefore this corresponds to a sugar translocation velocity of 62 cm/h.

Although this value would seem reasonable, at least when compared with the results of other workers using different techniques, the objection could still be raised that Zimmermann's reasoning was faulty. The concentration gradient down the tree could well have been the result of a gradient of negative pressure in the xylem; the drop in concentration consequent upon defoliation may have been caused by a reduction in xylem tension when movement of the transpiration stream ceased after leaf removal (Peel and Weatherley, 1962).

A much better approach to the measurement of velocity in the sieve tubes of American ash was made by Zimmermann in 1969. Zimmermann had already demonstrated (1957a) that the carbohydrate in exudates from this species was not just in the form of sucrose, but is a mixture of stachyose, raffinose, sucrose and D-mannitol. He also found that there was a small diurnal variation in the concentration ratios of these sugars, and proposed that velocities could be measured by tracking the 'ratio wave' of these sugars down the trunk of a tree. He argued that a moving wave of concentration ratios should be entirely independent of the absolute sugar concentrations, and thus be unaffected by changes in xylem pressures. Certainly, no evidence is available to suggest that a change in xylem tensions can affect a *ratio* of concentrations, even though absolute concentrations can be affected (Weatherley *et al.*, 1959).

Figure 4.1 presents the data from one of Zimmermann's 1969 experiments in which exudate samples were collected from three bark incisions made on three different sides of the stem at heights of 0.5, 3.5, 6.5, 9.5 and 12.5 m. From results such as these it was calculated that the curve of concentration ratio moved down the stem at a velocity of between 30 and 70 cm/h. It is of interest to note that the velocity appeared to be slightly greater in the upper than in the lower part of the tree.

Radiotracer methods

The usual experiment performed with radiotracers is to introduce labelled molecules into the transport system, usually via a leaf, and then to detect after a known interval of time the position of the 'front' of advancing radioactivity. While there is no doubt that if activity moves x cm in y h, the velocity of transport must be at least x/y cm/h, there are a number of reasons for believing that the values obtained may not give an accurate measure of the velocity of transport.

The first difficulty with such experiments is that no account is taken of the time necessary for the labelled substance to penetrate the leaf from application site to the transport system. If penetration takes place by diffusion, then a large proportion of the experimental time may have been taken up by molecules moving very short distances, e.g. through the leaf mesophyll. Thus the value obtained will be less than the actual velocity of movement.

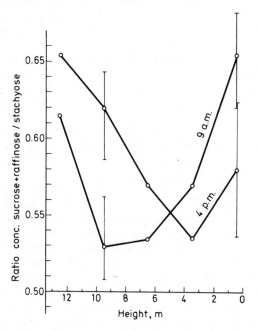

Figure 4.1 Gradients of sugar concentration ratios along the stem of Fraxinus americana. *The curve advances down the stem at a velocity of 30–70* cm/h. (*From Zimmermann, 1969, courtesy of Springer-Verlag*)

To some extent this problem can be removed by having two detectors of radioactivity on the stem of a plant, the time interval being measured for activity to move from the apical to the basal detector. Peel and Weatherley (1962) employed two colonies of aphids, sited a known distance apart on willow cuttings. $^{14}CO_2$ was applied to the leaves of the cutting, the arrival of activity at the site of each colony being found by collecting honeydew samples (*Figure 4.2*). The values obtained in these experiments lay between 25 and 33 cm/h. Clearly, here there could have been no effect on the velocity data of the time taken for uptake, assimilation and movement of the

[14]C-label into the sieve elements of the leaves, since the elapsed time was not measured until activity had reached the apical colony.

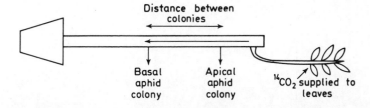

Figure 4.2 Experimental arrangement for measuring the velocity of [14]*C-labelled assimilate transport in willow stems, using colonies of* Tuberolachnus *to sample the sieve tube sap*

A second problem which arises in the measurement of velocities using radiotracers is that the measured distance the 'front' of radio-activity travels in a given time depends upon the sensitivity of detection. As Canny (1960b) has pointed out, 'as long as we envisage the translocated substance moving down the transport conduit as a piston travels down a tube, with a sharp demarcation at the front, the error involved is not likely to be very great'. However, the profile of a translocated substance does not show a steep change in concentration at a particular site in the phloem; the experimental measurement of translocation profiles shows them to have a logarithmic form (Vernon and Aronoff, 1952; Swanson and Whitney, 1953; O. Biddulph and Cory, 1957).

This effect of the sensitivity of detection on the measured distance of the front of radioactivity may be seen by reference to *Figure 4.3*. With a small dose of radioactivity applied to the system, the profile would be A, the measured distance of the 'front' being about 8 cm. With a higher dose of activity, the profile would be B, the measured distance increasing to 11 cm. The equation for the lines A and B is

$$\log r = \log a - bd$$

where r is the radioactivity at any point at a distance d from the site of application to which a dose of radioactivity equal to a was given, and b is a constant. The distance at which the smallest measurable quantity of radioactivity r_1 can be detected is

$$d = \frac{1}{b} \log \frac{a}{r_1}$$

From this expression it is clear that doubling the dose of radioactivity, or doubling the sensitivity of detection, will make the 'front' appear 30% further down the stem. To make the 'front' appear twice as far down the stem, the dose of radioactivity must be increased from a to a^2. Therefore the measured velocity of transport will depend upon the quantity of radioactivity applied at the source of the translocation system (Canny, 1960b).

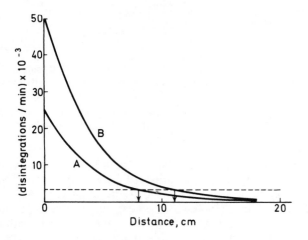

Figure 4.3 Graph on linear axes of radioactive concentration versus distance. The dotted line represents a hypothetical limit of detection of radioactivity and the intercepts of this line with the curves show the distances from the application site at which radioactivity would be detectable with the two doses of radioactive substance. (From Canny, 1960b, courtesy of Cambridge University Press)

If the application of the tracer is prolonged, it is possible that the advance of the front may be affected by the rate of entry of the tracer into the transport system. Canny (1960b) has quoted some data obtained by O.Biddulph and Cory (1957) and by Vernon and Aronoff (1952) to illustrate this point. With long tracer application times the measured velocities of transport would markedly increase. Therefore, tracers, if used to measure velocities, should be applied only as a short pulse to the translocation system.

A further difficulty encountered in the measurement of velocities using tracers is that associated with lateral loss from the transport conduits (*Figure 4.4*). Here we have two transport systems, each of which is transporting a labelled substance X from point A, the site of application, to B, the site of detection. In one system a high lateral leakage of X takes place; in the other system the lateral loss of X is lower. If the detection limit of X is the same in the two systems, it is evident that a lower velocity will be measured in the system with the high rate of lateral loss than in the system with a lower rate of lateral loss. It is also clear that this would occur even though the 'actual' velocity of movement of X in the two systems was the same.

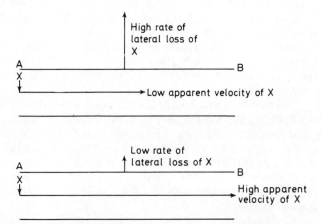

Figure 4.4 Effect of lateral loss of tracer from a transport system on the measured velocity of movement. (From Peel, 1972a, courtesy of Academic Press)

Experimental evidence of the effect of lateral loss of radioactivity on measured velocities has been given by Hoad and Peel (1965a). Using a two-aphid-colony system on willow cuttings (*Figure 4.2*), they were able to alter the rate of radial movement of ^{14}C-labelled assimilates from the phloem to the xylem by changing the carbohydrate status of the cuttings. The data presented in *Table 4.1* show that the rate of lateral loss from the sieve tubes had a profound effect on the measured velocities of [^{14}C]sugar transport. A comparable effect was found by Hoad and Peel, using ^{32}P-labelled phosphates.

Table 4.1 EFFECT OF LATERAL LOSS OF ^{14}C-LABELLED SUGARS FROM THE PHLOEM TO THE XYLEM ON THE MEASURED VELOCITIES OF TRANSPORT IN WILLOW CUTTINGS. (FROM HOAD AND PEEL 1965a, courtesy of *The Clarendon Press*)

Carbohydrate status	Lateral loss from sieve tubes	Measured velocity of [14]C-*sugar transport*, cm/h
High	Low	13.3
Low	High	6.6
High	Low	21.0
Low	High	3.5

10 μCi of [14]CO$_2$ applied to leaves of willow cutting in each experiment.

Some most interesting experiments in which velocities were measured using radiotracers were performed by Moorby, Ebert and Evans (1963), and by Evans, Ebert and Moorby (1963), on soyabean. Using the short-lived ^{11}C isotope, which emits high-energy radiation, instead of ^{14}C, these workers were able to study the movement of ^{11}C-labelled sugars *in vivo* after allowing the leaves to assimilate in $^{11}CO_2$. Moreover, they were able to perform a number of experiments on the same plant by allowing the ^{11}C to decay away between treatments.

They demonstrated that $^{11}CO_2$ did not move down the plants by gaseous diffusion but was transported in the form of labelled assimilates. The measured velocities of transport were 60 cm/h, and from a mathematical analysis of the data it was estimated that 0.8 % of the material in the sieve tubes leaked out per centimetre of stem traversed. In view of this low rate of loss, it must be assumed that the velocity measurements were unlikely to have been affected to any marked degree.

Physical methods

If movement of solutes in a transport system such as the phloem takes place by means of a bulk flow of solution, then it should be possible to measure the velocity of this solution by a thermoelectric technique. Huber *et al.* (1937), having already developed this technique for measuring the velocity of the transpiration stream, attempted to apply it to movement in the phloem. The principle of this method is that a small area of the conducting tissue is warmed in a position midway between two temperature detectors. The time taken for the heat pulse to reach each detector is then measured. If there is a sap stream moving in the direction of one detector, the speed of the pulse is increased compared with the speed at which the pulse reaches the second detector. In practice, Huber and his colleagues were unable to measure the velocity of phloem transport using this method on trees, owing to interference by the xylem stream moving in the opposite direction.

However, Ziegler and Vieweg (1961) have adapted the technique for use on herbaceous material. Experiments were performed on a species of *Heracleum* in which vascular bundles can be exposed while still attached to the plant. Moreover, it is possible to separate the xylem from the phloem over short distances, thereby eliminating any distortion caused by the transpiration stream. Using the thermoelectric technique on these exposed phloem tissues, Ziegler and Vieweg were able to measure velocities of movement in the phloem of 37–70 cm/h. Apart from the fact that this technique has many fewer deficiencies than radiotracer methods, thus giving more accurate estimates of velocity, the demonstration of heat transport in the phloem is cogent evidence for a bulk flow of solution.

Despite the difficulties encountered in velocity measurements, most workers in the field of phloem physiology would accept that solutes can be transported through the sieve tubes at speeds of up to 100 cm/h. Some experiments have given values greater than 100 cm/h, and it seems possible that in certain systems high velocities can be obtained. Nelson, Perkins and Gorham (1958) have claimed that ^{14}C-labelled assimilates can move at speeds up to 3 m/h in stems of soyabean. In contrast, Canny (1961), utilising the time and distance profiles of ^{14}C-labelled assimilates in *Salix* stems, has given values for the velocity of phloem transport as low as 1 cm/h.

The measurement of rates of mass transfer

While estimates of the velocity of phloem transport can differ by several orders of magnitude, many of the measurements of mass transfer rates give values which are remarkably similar in a range of species. Generally speaking, mass transfer rates are somewhat difficult to measure and it is not surprising, therefore, that the number of such measurements given in the literature is rather small. A number of workers have chosen organs such as tubers or fruits as convenient systems for the measurement of mass transfer rates through the phloem of the stem leading to the organ; others have determined rates through the phloem of petioles. A description of the experiments of Dixon and Ball (1922) has been given in Chapter 1 (p. 27). A list of the published values was given by Canny (1960b), reproduced here as *Table 4.2*.

These values have been reduced to the common units of grams dry weight per square centimetre of phloem per hour, the figure in parentheses after each value giving the number of measurements on which each is based. Canny has drawn attention to the similarity between the values from different stem systems and the similar agreement between the petiole systems, as being of great importance with regard to the mechanism responsible for movement.

If it is assumed that the sieve tubes occupy approximately one-fifth of the total area of the phloem, and measurements have been made which support this figure (Mason and Lewin, 1926; Crafts, 1931; Crafts and Lorenz, 1944), then the data can be expressed in units of grams per square centimetre of sieve tube per hour. A mean value for the stem systems would thus be 20 g cm^{-2} sieve tube h^{-1}. Other estimates of the proportion of sieve elements in the phloem have revealed that in certain trees the sieve elements can make up to 75% of the total cross-sectional area (Lawton and Canny, 1970).

Table 4.2 RATE OF TRANSLOCATION AS MEASURED BY MASS TRANSFER OF DRY WEIGHT. (FROM CANNY, 1960b, courtesy of the *Cambridge University Press*)

Plant system	Specific mass transfer, g dry wt. $cm^{-2}phloem \ h^{-1}$	Reference
A. STEMS		
Solanum tuber stem	4.5 (1)	Dixon and Ball (1922)
Dioscorea tuber stem	4.4 (1)	Mason and Lewin (1926)
Solanum tuber stem	2.1 (140)	Crafts (1933)
Kigelia fruit peduncle	2.6 (1)	Clements (1940)
Cucurbita fruit peduncle	3.3 (39)	Crafts and Lorenz (1944)
Cucurbita fruit peduncle	4.8 (5)	Colwell (cited in Crafts and Lorenz, 1944)
Gossypium bark flaps (probably damaged)	0.14–0.64 (9)	Mason and Maskell (1928b)
B. PETIOLES		
Phaseolus petiole	0.56 (1)	Birch-Hirschfield (1920)
Phaseolus petiole	0.7 (1)	Crafts (1931)
Tropaeolum petiole	0.7 (1)	Crafts (1931)

The accuracy of these measurements is important. Since all the evidence points to the sieve elements being the transport channel, any measurement of specific mass transfer rates must be on the basis of these cells. Certainly, more accurate determinations of mass transfer rates must be made on as wide a variety of plant systems as is possible. It would be particularly useful if these measurements could be correlated with growth rate measurements under well-defined environmental conditions of light, temperature, etc., and also with the stage of differentiation of the plant organ under investigation. This would provide us with many more data on the magnitude of the transport process. It would be particularly interesting to obtain confirmatory evidence of differences in mass transfer rates in different plant organs, as shown by the data for stems and petioles given in *Table 4.2*. The problem of the control of phloem transport will be considered in the following chapter, but, clearly, if the sieve tubes of different organs are found to conduct at different mass transfer rates, then this would constitute a fundamental problem which would have to be taken into account by any hypothesis of the mechanism of movement.

Before we leave the subject of the rate of translocation, it should be mentioned that Canny (1960b) has referred to the aphid stylet experiments of Weatherley *et al.* (1959) as the only instance in which mass transfer rates in a single sieve tube have been measured. An average figure of 1 µl/h was given for the volume flow rate from the severed stylets, the exudate having a concentration of 10% (w/v) of sucrose. The diameter of the sieve tubes in the willow used by Weatherley and his colleagues was 23 µm; therefore this would represent a specific mass transfer rate of 0.1 mg sucrose per hour per 414 µm², or 24 g sucrose per square centimetre of sieve tube per hour. Canny makes the point that this is in remarkable agreement with the other data for stem systems. This cannot be denied, but a word of caution should be added: we have yet no conclusive evidence that movement of solutes towards a sieve element pierced by a stylet only involves movement through a single sieve tube. There is a distinct possibility that more than one sieve tube may be participating in the stylet exudation process.

5

The control of the rate and direction of phloem transport

There seems little doubt that the goal of many phloem physiologists is the elucidation of the mechanism of longitudinal movement in sieve tubes. Praiseworthy as this aim is, it only constitutes one facet of the problem. Mechanisms are so fascinating a subject that they may have received undue attention; equally important aspects tend, with certain exceptions, to have been relegated to a somewhat lower plane. It must be clear, however, that to obtain a complete picture of phloem transport we must find out how the process is controlled. In some ways it is possibly more important to investigate control than cellular mechanisms; at least, this knowledge would enable us to manipulate movement for horticultural purposes. However, it is not suggested that studies of control and of mechanism are separate, alternative ways of studying translocation, for a true understanding of the process can only be attained by a synthesis of both methods of approach. For a comprehensive review of assimilate movement, the reader is referred to the article by Wardlaw (1968).

Any transport system must consist of three parts: (1) a source of solutes, (2) the transport conduits or path and (3) the sink at which the solutes are removed from the conduits. We can define a source as a tissue or organ in which the export of a given solute exceeds its import, the converse being true of a sink. The point must be made, however, that an organ could be a source for one particular solute and a sink for another solute, although whether these two states could exist concurrently is not known; it may be that they have to be separated in time, albeit by only minutes.

From the spatial relationship, source→conduit→sink, it is clear that each part of the system must play a role in the over-all control of transport. Solutes cannot be transported at a greater rate through the conduits than they can be loaded at the source; conversely, the rate of translocation will be dependent on the rate at which solutes are removed at the sink. The juxtaposition of sources and sinks should, therefore, clearly decide the direction of transport, although here again it is possible that the conduits might have an influence if their conductivity was greater in one direction than in the other.

The work described in the present chapter is largely concerned with source–sink interactions, the effect of their removal upon the rate and direction of transport, and environmental effects upon movement; i.e., in a sense, what could be termed 'natural' changes in the plant. Over the past decade, however, a considerable body of information has been produced which shows that the application of natural growth substances to plants can markedly affect the rate and direction of solute movement. Since it is not yet clear to what extent this 'hormone-directed transport' is related to conditions occurring in untreated plants, particularly since the hormones are often applied at high concentrations, a discussion of this work is not given in the present chapter. It was considered to be more appropriate to deal with hormone effects separately in the chapter in which hormone transport is also covered (Chapter 11, pp. 214).

Control by sources and sinks

As a continuation of· their work on the pathway of transport (1928a), Mason and Maskell (1928b) published a second paper in which they described experiments on the factors controlling the rate and direction of movement in cotton plants. They investigated this problem using two different methods. The first of these was an attempt to make a green, mature leaf import sugar by darkening it with a black paper bag. The results, however, proved inconclusive.

The second method gave indisputable evidence that the 'normal' direction of transport could be reversed by changing the relative positions of sources and sinks. Essentially, the procedure consisted of baring a region in the middle of a stem of all leaves and branches. In one group of plants the leaves above this region were removed, and in another the leaves below. The source of sugar supply was thus in one case above the sink, and in the other below it, the bared region in the middle providing the sink. The results showed conclusively that sugar could be transported both upwards and downwards, depending only on the position of the source of sugar supply.

Mason and Maskell concluded that sinks played a relatively minor role in the regulation of transport. While this has subsequently been shown to be not entirely correct, there is no doubt that both the activity and position of sources do play a major role in the control of transport. The following generalisations are often made with regard to leaves. The lower leaves of a plant export primarily to the roots and the upper leaves to the shoot apex, while intermediate leaves export in both directions. Young, growing leaves only import nutrients, while fully mature leaves only export assimilates, although mature leaves can apparently import small quantities of assimilates (Thaine, Ovenden and Turner, 1959). Thrower (1962) carried out a study of the changes in the rate of export of ^{14}C-labelled assimilates by expanding leaves of soyabean. No export of the label occurred when the leaf area was less than 30% of adult size; thereafter export increased, an overlap being found between 30 and 50% of the adult size where the leaf is simultaneously importing and exporting assimilate.

Surgical treatment, i.e. pruning or defoliation of a plant, can markedly change the distribution pattern of assimilates. Geiger and Swanson (1965a) pruned sugar beet plants so that export from a single, mature source leaf occurred mainly into a young, sink leaf. Forde (1966) used defoliation to alter the pattern of assimilate movement in grasses. Thrower (1962) has published results which show that defoliation between a source leaf and the root caused more ^{14}C-labelled assimilate to move to the root and less to the apex in soyabean (*Figure 5.1*).

Thaine *et al.* (1959) demonstrated that the removal of all the mature leaves between the source and the apex in soyabean significantly increased the movement of labelled assimilates upwards. As the leaves which were removed normally would have imported little, Thaine and his colleagues presumed that this enhanced upward movement was due to the lowering of the concentration of total assimilate at the apex, resulting from the removal of leaves which would normally have exported unlabelled assimilate. Shading of the green apex in some, but not all, experiments led to a decreased movement to the apex from the mature source leaf. Their general conclusions were that the direction of movement from a given leaf depends upon the age and position of the leaf on the stem.

Hartt, Kortschak and Burr (1964) studied the effects of defoliation in sugar cane on the distribution of labelled photosynthate. They found comparable effects to those described in soyabean by Thrower (1962). When all the leaves below the [14]CO_2-fed leaf were removed, there was a small decrease in the percentage of the label translocated from the fed leaf to the stem and leaves above, and a small increase in translocation to the roots. They found that defoliation of all but the fed leaf gave greater translocation than with no defoliation. This fact Hartt and her co-workers cited as being particularly interesting, since it indicated that transpiration by other leaves exerted no 'pull' on translocation. Some earlier work by Yang (1961) had indicated a possible relationship of transpiration to translocation in sugar cane, more translocate going to the actively transpiring parts than to the less active parts.

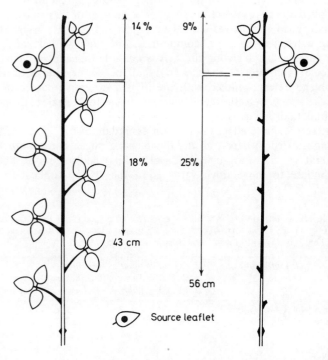

Figure 5.1 Distribution of [14]*C-labelled assimilate in defoliated and intact soyabean plants. (From Thrower, 1962, courtesy of C.S.I.R.O.)*

As previously mentioned, there is now ample evidence that the position and activity of sinks can play a considerable role in the control of transport. Wardlaw and Porter (1967) demonstrated that competition occurs between sinks. In mature plants of wheat which are bearing a heavy crop of seed, very little assimilate moves to the roots during seed development. Carr and Wardlaw (1965) followed the movement of labelled assimilates in wheat. For the first week after anthesis, half the assimilates leaving the flag leaf moved into the growing top internode. In low light, stem growth competed with the developing grain for assimilates. Nelson and Gorham (1957) demonstrated that sinks in soyabean plants could markedly affect the distribution of labelled photosynthate.

Khan and Sagar (1966), using tomato plants, found that assimilate distribution was rather different from that in many plants used for translocation work. In vegetative plants the lower leaves exported, more labelled carbon up than down, while the upper leaves moved a high proportion downwards. In young fruiting plants all the leaves supplied all the trusses, although, as the number of trusses increased, certain groups of leaves tended to supply individual trusses. Stems were active sinks for ^{14}C-compounds, although neither shoot tips nor roots acted as strong sinks. It is possible that the very complex vascular anatomy of tomato (we shall be considering anatomical problems later), coupled with the highly vegetative nature of this species, may explain the deviations from the usual patterns of assimilate distribution.

A very definite effect of sinks on assimilate distribution has been demonstrated by Hartt *et al.* (1964), using sugar cane, both lalas (lateral shoots) and suckers (basal shoots) causing more labelled assimilate to move downwards from the $^{14}CO_2$ application leaf (*Table 5.1*).

Table 5.1 EFFECT OF LALAS AND SUCKERS ON TRANSLOCATION IN SUGAR CANE. $^{14}CO_2$ (100 μCi) WAS FED TO A CENTRAL 20 cm LENGTH OF LEAF BLADE FOR 5 min AT 9000 ft cd. (FROM HARTT, KORTSCHAK AND BURR, 1964)

| *Part* | *Distribution of radioactivity in entire plant as percentage of RTC** | | | |
	Suckers 0 *Lalas* 0	*Suckers* 0 *Lalas* +	*Suckers* + *Lalas* 0	*Suckers* + *Lalas* +
Upper leaves and joints	1.9	3.9	3.3	4.7
Fed leaf	83.6	61.9	53.7	51.5
Stalk below fed leaf	14.4	32.3	40.8	41.1
Lalas	—	0.3	—	0.7
Suckers	—	—	1.8	1.2
Roots	0	1.9	0.3	1.5
RTC × 10^6	42	49	75	87

* Relative total counts, equivalent to relative specific activity times total dry weight in milligrams.

Feeding aphids act as sinks for assimilates. Peel and Ho (1970) have investigated the effect of colony size (i.e. sink activity) of *Tuberolachnus salignus* on the distribution of labelled assimilates in willow. The technique used by Peel and Ho is illustrated in *Figure 5.2*. Labelled carbon dioxide was supplied to the leaves of the willow cutting, the relative mass transfer rates of the labelled sugars to each colony being measured by determining the specific activity of the honeydew collected over a period of 15 h from the application of the tracer. Data from these experiments are shown in *Table 5.2.*

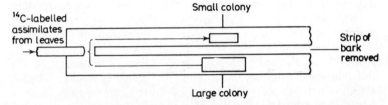

Figure 5.2 Experimental arrangement used to investigate the effect of aphid colony size on the movement of ^{14}C-labelled assimilate in willow. (After Peel and Ho, 1970)

Not only was the total activity produced by the large colony greater than that from the small colony but the specific activity of the honeydew from the large colony was also greater than from the small. Peel and Ho concluded that this was due to a greater contributory length of the large colony, i.e. it was receiving assimilates from further away than the small colony.

Table 5.2 EFFECT OF APHID COLONY SIZE ON THE SPECIFIC ACTIVITY OF HONEYDEW PRODUCED BY *Tuberolachnus salignus* (GMELIN). (FROM PEEL AND HO, 1970)

Experiment	Mean rate of honeydew production, mg *dry wt.*/h		Specific activity of honeydew, counts min^{-1} mg^{-1}	
	Large colony	*Small colony*	*Large colony*	*Small colony*
1	1.27	0.40	46 200	24 200
2	0.61	0.29	24 900	13 100
3	0.86	0.18	449 500	285 000
4	0.62	0.19	69 000	4 000
5	0.32	0.18	26 400	8

Anatomical considerations

If the matter is given a little thought, it becomes clear that not only must the relative positions and activities of sources and sinks control transport, but the arrangement of the vascular tissue between them must also exert some control. O. Biddulph and Cory (1965) studied movement from *Phaseolus* leaves by combining an autoradiographic technique with a fluorescence method for locating phloem bundles. Labelled assimilates moved downwards one node from the $^{14}CO_2$ application leaf, an anastomosis of bundles at the node permitting the assimilates to divide into upward- and downward-moving components. The upward-moving component was limited to bundles which alternated with those conducting the downward-moving component from the next higher leaf. Assimilates from leaves just beginning to export left the leaf traces at the insertion node and then moved directly upwards. A downward flow developed as the leaf matured.

Joy (1964) showed that phyllotaxis was very important in the distribution of assimilates in *Beta*. In untreated plants labelled assimilates moved only into leaves having a direct vascular connection with the source. Removal of mature leaves from the untreated side of the plant did, however, produce a transfer of label to the young sink leaves on the defoliated side. This movement presumably was able to take place by virtue of the complex vascular anatomy of the crown. Nelson and Gorham (1957) and S. F. Biddulph *et al.* (1958) have demonstrated that export of labelled solutes from a leaf is restricted to the same side of the stem as the source leaf. The translocation of assimilates from leaves of the same orthostichy to a particular file of seeds in the developing head of sunflower has been reported (Prokofyev, Zhadanova and Sobolev, 1957). Such a regularity of assimilate transport has been investigated in relation to phyllotaxis in tobacco (Jones, Martin and Porter, 1959; Shiroya *et al.*, 1961) and in cotton (Ting, 1963).

Ho and Peel (1969a) investigated the movement of labelled assimilates and phosphates between the leaves of willow in relation to the phyllotactic configuration. The leaves in *Salix viminalis* are arranged in five orthostichies (*Figure 5.3a*). The leaves on three of these orthostichies were found to share one main transport channel, a second channel being shared by the leaves on the other two orthostichies. Transport was shown to occur readily between the leaves on two orthostichies if these were separated by an angular distance of 72° (*Figure 5.3c*).

Control by the transport conduits

In considering the possible role of the sieve tubes in the control of the rate and direction of transport, we enter a field in which patent lack of knowledge and speculation are rife. This is not really surprising, since control of movement by the sieve tubes must inevitably be bound closely to the mechanism of transport, a subject on which, as we shall see later very little undisputed evidence is available. Topics such as simultaneous bidirectional movement and the problem of whether different solutes can move with different velocities are better left until a discussion of hypotheses of the mechanism has been undertaken.

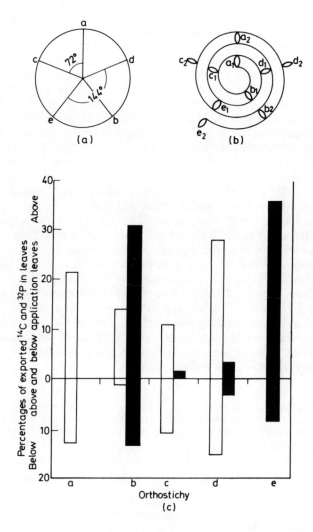

Figure 5.3 Diagrams illustrating the phyllotactic configuration in Salix viminalis. *(a) Angles and orthostichies. (b) Nomenclature of leaves. (c) Patterns of ethanol-soluble activity in leaves above and below application leaf (^{14}C to a_6, ^{32}P to b_6). Open columns, ^{14}C; closed columns, ^{32}P. (From Ho and Peel, 1969a, courtesy of The Clarendon Press)*

However, there are two aspects of the control of transport by the sieve tubes which can be profitably considered at this juncture. When mass transfer rates were discussed in the previous chapter, it was pointed out that we do not know whether the difference in the specific mass transfer data for stems and petioles shown in *Table 4.2* (p. 78) is a 'true' difference, or whether this merely reflects the source activity of the system. In a translocation system the mass transfer rate must be dependent upon the dimensions of at least two processes, viz. the rate at which solutes are loaded at the source and the rate at which they can be transported through the conduits. It is not known whether the situation could arise in which a source would have the potential to load solutes into the sieve elements at a faster rate than the sieve tubes could transport these away from the source, i.e. whether the sieve tubes can limit the rate of transport. It is difficult to conceive of an experimental technique which could give useful data on this question, but the problem is of considerable importance, not only in the context of the control of transport but also in relation to the mechanism of transport.

The second aspect of control by the sieve tubes is the problem of polarity, i.e. whether the conductivity of a sieve tube is greater in one direction than in the opposite. Some evidence is available which suggests that IAA movement may occur in sieve tubes in a polar fashion. This topic will be dealt with in a later chapter, but, as far as sucrose is concerned, there is no evidence that the sieve tubes exhibit polarity. Weatherley *et al.* (1959) investigated the problems in segments of willow stem by siting aphid stylets at the morphological base and the morphological apex. A comparison of the volume flow rates and sucrose concentrations from stylets at each end of the segment revealed no significant differences (*Table 5.3*).

Table 5.3 COMPARISON OF THE VOLUME FLOW RATE OF EXUDATION AND SUCROSE CONCENTRATIONS FROM APHID STYLETS SITED AT THE APICAL AND BASAL ENDS OF ISOLATED SEGMENTS OF WILLOW STEM. (FROM WEATHERLEY *et al.*, 1959, courtesy of *The Clarendon Press*)

Segment	Mean rate of exudation, $\mu l/h$		Difference	Least significant difference	Sucrose conc., $\mu g/\mu l$		Difference	Least significant difference
	Apical	Basal			Apical	Basal		
1	1.21	1.19	0.02	0.82	130	126	4	33
2	1.74	1.99	0.25	0.77	135	133	2	41

The effect of environmental factors on translocation

There are three environmental factors—water potential, light and temperature—which have been shown to affect the rate and patterns of translocation. From what has already been said in this chapter, it is evident that these factors could change the over-all rate of the transport process, either by acting on the source, the sink or the transport conduits, or by affecting all three parts of the system.

Light

There would seem to be at least two documented effects of light on the rate of translocation: one associated with the rate of photosynthesis, i.e. with the 'activity' of the source, the second with a possible photocontrol of translocation, dependent upon the quality of the light received by the plant.

It has been known for a considerable time that the 'activity' of a photosynthesising source leaf can be affected by light intensity, i.e. by the quantity of light energy given. Rohrbaugh and Rice (1949) showed that 2,4-D moved out of photosynthesising leaves, although little movement occurred in darkness unless sugars were also supplied. Other experiments with 2,4,5-T by Brady (1969) demonstrated that light intensity could have effects on both the uptake and export of this substance by leaves of *Quercus*.

A number of studies (Geiger and Swanson, 1965b; Hartt *et al.*, 1964; Moorby *et al.*, 1963) have shown that export of sugars from the source leaf declined rapidly after darkening, reaching a rate approximately 25% of the rate in light after 150 min of darkness. The data presented in *Table 5.4* were obtained by Thrower (1962), working on soyabean. These results demonstrate not only that the total activity exported by the uppermost expanded leaf was dependent upon light intensity but that the ratio of activity transported downwards (*B*) to that moved upwards (*A*) was also affected. Since the total activity in the plants was not significantly different between the two treatments, Thrower concluded that at both light intensities the leaf assimilated all the $^{14}CO_2$ supplied.

Thrower also demonstrated that pre-illumination of plants prior to the application of $^{14}CO_2$ could result in a reduced transport of the ^{14}C-label out of the leaf. In contrast, Nelson (1963) showed that more activity moved to the roots in seedlings of *Pinus strobus* which had been grown at low light intensities than in those grown at high light intensities. It seems reasonably certain that the effects of light intensity on the rate of translocation depends upon the 'physiological state' of the sources and sinks, i.e. on their carbohydrate status. In this connection, Geiger and Batey (1967) have demonstrated that polysaccharide reserves in darkened leaves of *Beta* begin to contribute to the translocate stream after 2–3 h of darkness.

Table 5.4 DISTRIBUTION OF LABELLED ASSIMILATE IN SOYABEAN PLANTS AFTER EXPOSURE TO DIFFERENT LIGHT INTENSITIES FOR 2 h. 5μCi $^{14}CO_2$ ADMINISTERED TO BOTH GROUPS. (FROM THROWER, 1962, courtesy of *C.S.I.R.O.*)

	High light intensity (1000–2000 ft cd)	*Low light intensity* (500–700 ft cd)
$10^{-4} \times$ total activity in plants	105.5 ± 10.7	97.7 ± 5.0
Percentage of total activity:		
above source leaf (*A*)	7.8 ± 0.5	3.4 ± 0.5
below source leaf (*B*)	36.3 ± 2.0	23.3 ± 1.9
which has moved out of source leaf	44.1 ± 1.9	26.9 ± 2.1
Distribution ratio (*B/A*)	4.7 ± 0.6	8.1 ± 1.1

From the results of experiments on the translocation of labelled assimilates in detached leaf blades of sugar cane, Hartt (1965b) concluded that the initiation of sugar translocation from the leaf is under photocontrol. This inference was based largely on the observation that photosynthesis in sugar cane was saturated at 6000 ft cd, while compensation was reached at about 125 ft cd. Since light affected the polarity of translocation at intensities below the compensation value, it was argued that this could not be merely an effect of light on the rate of photosynthesis.

In a further paper Hartt (1966) studied the effect of light of different wavelengths upon sugar transport in blades of sugar cane. Basipetal transport was stimulated in red or blue light more than in green or cool-white fluorescent illumination. Illumination by far-red light did not stimulate transport, but acted like total darkness. Because of the wide emission characteristics of the lamps she employed, Hartt could not conclude which pigment system was involved in the light stimulation of transport.

Water potential

The process which could certainly be affected by water stress is photosynthesis and, hence, the rate of loading of carbohydrates into the phloem by source leaves. Working with wheat, Wardlaw (1967) showed that the rate of photosynthesis of plants grown under water stress was reduced as compared with plants given adequate moisture. Although the rate of photosynthesis was lowered, the growth rate of the grain was not reduced, a greater proportion of the total assimilates moving upwards in the water-stressed plants. A similar effect of water stress on the distribution of labelled assimilates has been reported by Wardlaw (1969), using *Lolium*.

An effect of water stress on the distribution pattern of ^{14}C-labelled assimilates in saplings of yellow poplar has been described by Roberts (1964). In stressed plants the percentage of the total ^{14}C which was translocated was reduced. However, as with wheat and *Lolium*, an increased percentage of the total activity was transported upwards. Roberts attributed the reduced translocation in stressed plants to a lowering of photosynthesis consequent upon stomatal closure.

Plaut and Reinhold (1965) applied ^{14}C-labelled sucrose to the lower epidermis of bean leaves, and studied translocation of the label in stressed and unstressed plants. The unstressed control plants generally translocated better than the stressed plants, both up and down the stalk from the application leaf. Translocation of ^{14}C to the roots was enhanced by an adequate moisture supply. Zholkevich and Koretskaya (1959) showed that drought conditions led to a reduced accumulation of sugars in pumpkin roots.

Some evidence that water stress may have a direct effect upon the rate of translocation, rather than an indirect effect caused by a reduction in photosynthesis, has been produced by Hartt (1967). Low moisture supply to sugar cane plants was effected either by adding sodium chloride to the nutrient solution bathing the roots, allowing a cut stalk to wilt, or withholding water from plants growing in the field. Plants in which water stress was induced had a lower relative rate of movement of [^{14}C] labelled assimilates than had control plants. Moreover, low moisture supply depressed the rate of translocation relatively more than it reduced photosynthesis. *Figure 5.4* presents some of Hartt's data, which show that, while photosynthesis only fell by 18 % in the stressed as compared with the control plants, translocation of [^{14}C]sugars in the stressed plants dropped by over 90 %. This led Hartt to conclude that the reduction of transport in the stressed plants must have been due to a direct effect upon the longitudinal transport of materials in the phloem, rather than to an effect upon the availability of assimilates in the source leaves.

A demonstration that water stress can directly affect translocation processes has been made by Weatherley *et al.* (1959), using the aphid stylet technique on isolated segments or bark strips of willow. Reduction in water potential was achieved by introducing

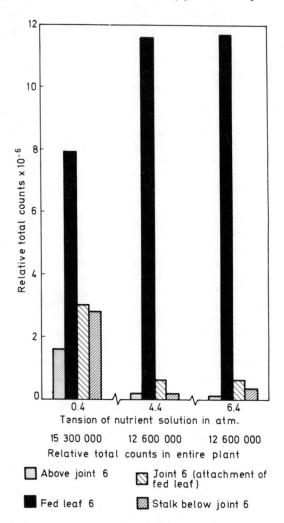

Figure 5.4 Effect of moisture supply on translocation in sugar cane. Plants were placed in their respective solutions ($\pm$ NaCl) at 10 a.m. and harvested 48 h later. Ninety minutes before harvest $^{14}CO_2$ was administered to blade 6 for 5 min in sunlight. (From Hartt, 1967, courtesy of the American Society of Plant Physiologists)

mannitol solutions into the xylem of segments, or on to the cambial surface of bark strips. Decrease in water potential led to a fall in the volume flow rate of exudation, a rise in the concentration of the exudate and a concomitant fall in the rate of sucrose exudation.

Weatherley and his co-workers suggested that the fall in the sucrose exudation rate was produced by a decrease in the water flux into the pierced sieve tube, which in turn led to a fall in the sucrose 'potential' gradient between the source cells of the bark and the sieve tube. This fall in potential thus produced a decline in the rate of sucrose loading.

An effect of moisture levels on acid invertase content of immature storage tissues of sugar cane has been demonstrated by Hatch and Glasziou (1963), low moisture regimes drastically reducing the level of this enzyme by a factor of up to 10. Since the level of this enzyme may affect the movement of sucrose from the phloem into immature tissues, i.e. into sinks, it is clear that moisture could have a considerable effect upon the partitioning of assimilates in sugar cane.

There seems little doubt that water stress can markedly affect translocation, in terms of either the rate of the process or the pattern of assimilate distribution. Most workers have shown that the rate of transport tends to be reduced by low moisture content, although there are isolated reports of the enhancement of transport. Eaton and Ergle (1948), working on cotton, and Ehara and Sekioka (1962), using sweet potatoes, have concluded that translocation of carbohydrates to the root system was stimulated by low relative humidity or soil moisture.

From the work which has been quoted, it seems clear that water stress may influence translocation by acting upon either the loading of solutes at the source or the unloading at the sink tissues. We do not yet have any undisputed evidence that a lowering of water potential can directly affect transport within the sieve tubes. If transport is accomplished by a mass flow mechanism, it is, however, very easy to envisage a retardation of transport at low water contents due to an increase in the viscosity of the moving solution. Even a mechanism in which solute molecules moved through a static water phase might be subject to adverse effects of low water potential owing to dehydration of the sieve elements.

Temperature

Probably no environmental factor has received more attention than has temperature. However, the majority of experimenters have directed their efforts towards an elucidation of the effects of temperature (particularly low temperatures in the region 0–5°C) on movement through the pathway, i.e. the phloem. To explain this bias requires a discussion of the energy requirements of the various hypotheses of the mechanism of transport; therefore it seems preferable to postpone consideration of localised path chilling experiments until a review of the hypotheses has been given. At this point the review will be confined to work on temperature effects on the source and sink ends of the system, and on whole plants.

Experiments in which rates of translocation have been measured in whole plants maintained at different temperatures have revealed that the rate responds to temperature in a manner characteristic of an energy-dependent process, with a Q_{10} in the range 2–3, where Q_{10} is the temperature coefficient, i.e. the ratio

$$\frac{\text{rate of process at } T°C + 10°C}{\text{rate of process at } T°C}$$

Hewitt and Curtis (1948) carried out work on temperature effects on respiration and translocation from leaves of bean, milkweed (*Asclepia syriaca* L.) and tomato. Losses of carbohydrates due to respiration were measured on detached leaves, these, together with the plants on which translocation losses were measured, being kept in darkness during the experimental period. Translocation from leaves was greatest at 20 and 30°C and lowest at 4 and 10°C, while 40°C gave an intermediate value.

Thrower (1965) studied the effects of low temperature (2–3°C) on transport out of soyabean leaves. She 'spot' fed $^{14}CO_2$ to the terminal leaflet of the uppermost expanded leaf and after 3 h harvested the plants. Low temperature, although it did not markedly affect $^{14}CO_2$ uptake, completely inhibited the movement of activity out of the application leaf. A study of the translocation profiles of ^{14}C in *Pteridium* plants by Whittle (1964) led this worker to the conclusion that the optimum temperature for the process lay between 25 and 30°C. The Q_{10} of transport was found to be 2.9.

While work concerning temperature effects on transport in which the whole plant is subjected to a given temperature undoubtedly has some value, such experiments do not provide any clear information as to the responses of the various parts of the transport system. In the work quoted so far, the effect of temperature might well have been due to change in the rate of solute loading in the source leaf. Hartt (1965a) studied temperature effects on transport in sugar cane, using a method in which the temperature of the leaves and of the roots could be regulated separately. A drop in leaf temperature from 34 to 20°C produced a considerable fall in the amount of [^{14}C] assimilates exported by the $^{14}CO_2$ application leaf.

Hartt also demonstrated an effect of air temperature on the distribution of the labelled assimilates; at 34°C the ratio, activity moving up the stem/activity moving down, was greater than at 20°C. Also, the temperature coefficient of acropetal movement was higher than for basipetal transport (*Table 5.5*). It seems likely that

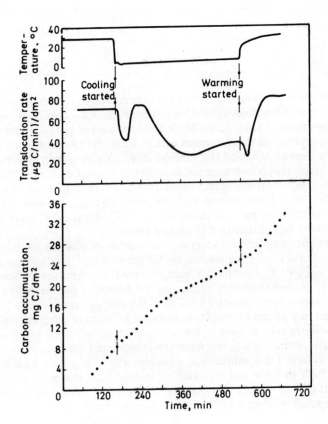

Figure 5.5 Rate of translocation of labelled photosynthate during cooling and subsequent warming of the entire sink region in sugar beet (roots, beet, crown and sink leaf). Labelling with $^{14}CO_2$ began at zero minutes. (From Geiger, 1966, courtesy of the American Society of Plant Physiologists)

these two effects are correlated; they could both be the result of a reduction in the rate of enzymic processes at the growing point sink.

Lowering of root temperature decreased translocation from the fed leaf only at high light intensities. Hartt concluded that this could have been caused by increased moisture stress in the cold-root plants.

Table 5.5 TEMPERATURE COEFFICIENTS OF ACROPETAL AND BASIPETAL TRANSPORT IN SUGAR CANE. (FROM HARTT, 1965a)

	Out of fed leaf	Translocation Basipetal	Acropetal
Q_{10} range, 24–34°C	1.1	1.05	3.9
Q_{10} range, 20–30°C	1.5	1.7	16.2

Fujiwara and Suzuki (1961) have carried out similar experiments on the translocation of [^{14}C]assimilates in *Hordeum*. They used temperatures of from 15 to 30°C around the aerial portions and the roots, and found the maximum rate of transport to occur when the leaves were at 25°C and the roots at 30°C. In view of what has beer previously said concerning the activities of sources and sinks in the control of transport rates, this is not a surprising result. These workers also demonstrated that photosynthesis in *Hordeum* unde. their conditions had an optimum of 25°C, while 30°C gave the maximum respiration rate in excised roots.

The effects of sink cooling on translocation of labelled assimilates in sugar beet has been studied by Geiger (1966). Upon cooling the sink leaf to 1°C, four phases were observed; a temporary decline, a period at the pre-cooling rate, a second period of decline, followed by a new steady rate at 35–45% of the original rate (*Figure 5.5*). Geiger interpreted his results as showing an effect of temperature on an active uptake process in the sink regions.

There seems little doubt from the experiments we have examined in this chapter that sources and sinks play a very large part in controlling both the rate and direction of transport. It is clearly possible drastically to change the pattern of transport, either by surgical manipulation or by altering environmental conditions.

6

The structure of phloem cells

Up to this point, with the exception of the section on ion movement across the root cortex in Chapter 1, we have been considering aspects of solute transport in which there is a fair degree of unanimity in the interpretation of the experimental data. Even though there is a divergence of opinion as to the relative merits of velocity and mass transfer as useful parameters of phloem transport, at least most workers on phloem would agree with the values for these parameters quoted in Chapter 4.

Now, however, we enter an area where the available data are subject to considerable controversy as to their interpretation. This area is, of course, that concerned with the mechanism responsible for movement in sieve tubes, and the following five chapters will to all intents and purposes be devoted to a discussion of this problem.

The purpose of the present chapter is to provide a review of investigations with the optical and electron microscopes into the structure of phloem cells. This subject is so intimately bound up with the possible mechanisms of movement that it seems inconceivable that the mechanism can be elucidated until the structure of the functioning sieve element is understood. Indeed, when we come to consider the hypothesis of the mechanism of transport, we shall see that each relies on a particular type of structure being present in the sieve element, and therefore any may be discarded if the required structure is not realised. For an excellent account of the relationships between structure and function, the reader is referred to the review by Weatherley and Johnson (1968).

The type of structure assigned to the presumed mature sieve element can vary widely between different workers. This variability in the interpretation of structure has frequently been so pronounced

97

that it must often have appeared to the outside observer that some workers were investigating quite different cells from others. Thus it is quite possible to choose a particular hypothesis, then subsequently scan the literature and find an account of the required structure.

The ontogeny of phloem cells

Angiosperms

The literature on the ontogeny and differentiation of phloem cells using both optical and electron microscopes is very extensive, and it is not the purpose here to give the reader more than a brief outline of the situation. Those who are particularly interested in this aspect of the subject are referred to two works by Esau (1965, 1969), the latter being a particularly comprehensive survey, not only of phloem differentiation but also of the structure and function of the mature cells.

Evert *et al.* (1969) have reported a study with the optical microscope of phloem differentiation in *Ulmus americana*, using material which had either been chemically fixed prior to sectioning, or freeze substituted and then fixed, or fresh samples which were plunged into 0.25M sucrose immediately after being removed from the tree.

Evert and his colleagues confirmed that both the sieve elements and companion cells arose from a single cambial mother cell. After the final division of the mother cell, the immature sieve element undergoes rapid expansion both radially and tangentially, the degree of expansion depending to a large extent upon the position of the element in the current year's growth; early sieve elements expand most, late elements least. Concurrently with this expansion process, the sieve element protoplast becomes highly vacuolated, so that in the fully expanded cell there is a large central vacuole surrounded by a narrower layer of cytoplasm.

During the expansion of the sieve element, three processes occur which eventually lead to the formation of the mature cell, viz. slime body formation and dispersal, nuclear degeneration and the formation of the sieve plates on the end walls of the cells.

Slime bodies, generally ovoid in shape, appear in the cytoplasm of differentiating sieve elements (Esau and Cheadle, 1965). After the first-formed slime bodies have increased in size, numerous other slime bodies appear in the parietal cytoplasm. Both groups of slime bodies initially possess an amorphous appearance in *Ulmus* (Evert *et al.*, 1969). Later, depending upon the fixation process employed, some of the slime bodies assume a fibrillar or stranded appearance. Cronshaw and Esau (1968a) have demonstrated the existence of two distinct types of slime bodies in *Cucurbita* sieve elements: a large type which arises as fine fibrils and a smaller type which begins as groups of tubules. As slime bodies increase in size, their limits become less well defined, they fuse and eventually they become completely dispersed (Esau and Cheadle, 1965).

During slime body fusion and dispersal, the nucleus starts to degenerate, loses its chromaticity and frequently becomes difficult to discern, particularly when it is masked by slime. Esau and Cheadle (1965) have given an account of nuclear breakdown in sieve elements of *Cucurbita*. In young sieve elements the chromatin units are large and stain well with Feulgen. However, when the slime bodies begin to fuse, the chromatin units become smaller and gradually disappear. The nucleolus also disappears, although a small spherical body may remain in the nucleus until the last stages of differentiation.

It appears that most sieve elements as they reach maturity lose the nucleus altogether, although some investigators have suggested that degenerate nuclei may remain in mature elements of certain plants, e.g. conifers (Murmanis and Evert, 1966; Wooding, 1966) and certain palms (Parthasarathy, 1966, quoted by Evert *et al.*, 1970). More recently, Evert *et al.* (1970) have reported nuclei of normal appearance in apparently mature sieve cells of certain gymnosperms and in some woody angiosperms (*Robinia*, *Ulmus* and *Vitis*).

As well as slime body formation and dispersal and nuclear degeneration, a third most significant process occurs during sieve element expansion—the formation of the sieve plates. Esau, Cheadle and Risley (1962) have investigated sieve plate formation in *Robinia* and *Cucurbita*. The sites of the future pores are first delimited by the appearance of small deposits of callose in the form of platelets. These callose platelets in *Ulmus* are at first narrower than the future pores, but soon increase in size and assume the angular shapes of the fully formed pores (Evert *et al.*, 1969). Endoplasmic reticulum becomes applied to the pore sites in *Robinia* and *Cucurbita* (Esau *et al.*, 1962) and is associated with the callose until the pores are finally formed. The actual perforation of the wall occurs in the centre of a platelet in such a way that the middle lamella disappears, the callose platelets fuse and a break occurs in the fused part, the process thus forming the pore, which is lined by callose from its inception. A very similar picture has been described by Wark and Chambers (1965), working on *Pisum*.

The fully differentiated sieve elements of angiosperms are generally 20–40 μm in diameter, although in some species they can have a diameter approaching 100 μm. The length of the cell usually lies between 100 and 500 µm. Figures in the range 0.1 to 5.0 µm are often quoted for the diameter of the pores in mature sieve plates.

Gymnosperms

The phloem of gymnosperms, although it differs from that of angiosperms in several important respects, appears to possess a fundamentally similar type of organisation. The conducting cells of gymnosperms are usually termed sieve cells rather than sieve elements. They do not have highly developed sieve plates on the end walls, connections between adjacent cells being confined to sieve areas on the lateral walls. Companion cells are not found, but it is generally assumed that their role is taken over by the so-called albuminous cells.

The differentiation of gymnosperm phloem has been described by Kollmann (1964), Evert and Alfieri (1965) and Wooding (1966), in *Metasequoia* and certain of the Pinaceae. The sieve cells arise from cambial derivatives which have a thin, parietal layer of cytoplasm containing mitochondria, plastids and a granular nucleus. As the young sieve cells expand, the plastids become localised at the position of the future sieve areas, and bodies which can be interpreted as slime bodies appear in the cytoplasm. These bodies then begin to elongate, forming strands, and at the same time the nucleus loses much of its chromaticity. Eventually the slime bodies disperse, the plasmodesmata of the primary pit fields become converted to sieve area pores and the tonoplast disappears (Evert and Alfieri, 1965).

The albuminous cells of *Pinus* are found in both axial and ray systems (Srivastava, 1963; Alfieri and Evert, 1968). They contain nuclei, plastids, mitochondria and lipid bodies. At certain times of the year they may contain starch. It has been suggested by Murmanis and Evert (1966) that slime bodies form in albuminous cells and that these elongate at the same time as slime body expansion occurs in contiguous sieve cells.

Albuminous cells do not arise from the same mother cells as the sieve cells, but they appear to come into very close association with the latter in mature gymnosperm phloem. In *Pinus*, for instance, the sieve cells are linked to only one other type of cell, i.e. the albuminous cell (Wooding, 1966). According to this worker, the connection between the sieve and albuminous cells is a compound pore which appears to be very similar to those found between sieve elements and companion cells in *Acer* by Wooding and Northcote (1965).

The developmental stage at which sieve elements translocate

If it is hoped to try and relate the structure of sieve elements to their role in the movement of solutes, then it is essential that studies be made on cells which are patently functional in translocation. Until fairly recently, it seems to have been assumed that the sieve element became 'mature', i.e. functional, when the sieve plates had differentiated and when much of the contents of the cells had been lost.

However, there have been suggestions, mainly supported by a number of workers in Germany, that it is the younger sieve elements, still containing large quantities of cytoplasmic material, which are functional in solute transport. This view was initiated by the work of Schumacher (1933) from his observations on the transport of dyes, and has more recently been supported by the investigations of Kollmann (1965) and his colleagues. Wark and Chambers (1965), working on *Pisum*, have also reported that perforation of the sieve plates occurs when the nucleus and tonoplast of the cell are still intact, i.e. the sieve elements of this species appear to be in a position to translocate even though they still contain considerable amounts of protoplasmic material.

Kollmann (1965) tried to localise the functional sieve cells in branches of *Metasequoia* taken in October, by a combination of autoradiographic and aphid-feeding techniques. After the application of $^{14}CO_2$ to the leaves, radioactivity was only found in the youngest sieve cells adjacent to the cambium. Furthermore, Kollmann demonstrated that the stylets of *Cinara laricola* pierced the sieve cells associated with the first band of bast fibres.

Kollmann and Dörr (1966) continued the studies on the localisation of the translocation process in gymnosperms, using a species of aphid which feeds on juniper. They used plants obtained at two different times of year, April and August. With April plants exudation from severed stylets was sporadic, but this became more reliable when the August plants were used. Aphids which fed on young shoots pierced the outermost cells of the metaphloem, while those which were allowed to feed on older branches positioned their stylet tips in the youngest sieve cells between the cambium and the first band of fibres.

Kollmann (1967), using a histoautoradiographic technique which he says precludes any possibility of the movement of water-soluble assimilates after sections have been taken, has further examined the situation in *Metasequoia*. Once again using material collected in October, he has demonstrated that the youngest sieve cells apparently participate to the greatest degree in the movement of assimilates.

The conclusions reached by Kollmann from work on material collected late in the growing season have been criticised by Crafts and Crisp (1971). According to these authors, it is probable that the last-formed layer of sieve cells are the only ones remaining functional in October, since all the older sieve cells are by then nearing obliteration.

Certainly, despite the findings of Kollmann, there is a considerable weight of evidence to support the idea that sieve elements can continue functioning when they are several years old. Esau (1948) has documented the processes of dormancy and reactivation of sieve elements in *Vitis vinifera*. Elements going into dormancy develop massive amounts of callose on sieve areas in all stages of differentiation. During reactivation in spring, the callose diminishes in quantity and sieve areas assume the same stage of differentiation which they had attained before the onset of dormancy. Davis and Evert (1970) have studied the seasonal cycle of phloem development in woody vines. Differentiation of phloem is completed by early August and cessation of phloem function begins in October. In *Vitis riparia* sieve elements usually function for more than one year, reactivation of the overwintering elements beginning in April.

Heyser, Eschrich and Evert (1969) have pointed out that the concept of the short life of sieve elements mainly comes from studies on deciduous dicotyledons. However, they quote work by Evert (1962) which indicates that the sieve elements of *Tilia* remain living, and therefore presumably functional, for periods of between 5 and 10 years. Furthermore, Evert *et al.* (1968) have demonstrated that two-year-old elements in *Tilia americana* were functional, since aphids were able to feed on these cells.

Heyser and his colleagues draw attention to the lack of data on the situation in perennial monocotyledons, many of which lack secondary vascular tissues; therefore it could be that some sieve elements remain functional for the life of the plant parts in which they occur. They quote work by Parthasarathy and Tomlinson (1967), who have demonstrated living sieve tubes in *Sabal palmetto* at least 50 years old, and the estimates of Tomlinson (1964) that in certain arborescent monocotyledons the conducting tissues must be more than 100 years old, to support their thesis on the longevity of the sieve elements of certain species.

In order to prove that relatively old sieve elements are capable of conduction, Heyser *et al.* (1969) applied ^{14}C-labelled phenyl alanine to exporting leaves on 30-month-old plants of the perennial monocotyledon *Tradescantia albiflora*. Histoautoradiographic examination of the basal internodes showed that radioactivity was present in the metaphloem of the bundles.

It seems possible to conclude from the evidence we have that sieve elements in a number of species can continue to function for a number of years, i.e. for a considerable time after they have become 'cytologically mature'. Certainly, it is essential before drawing inferences as to the structure of functioning sieve elements to determine whether the particular cells under observation are indeed capable of transporting materials. The efforts of cytologists in this difficult task should be appreciated and encouragement given to them to continue this work.

The ultrastructure of phloem

Companion and parenchyma cells

Esau and Cheadle (1965) have given a description of the ultrastructure of companion and parenchyma cells in *Cucurbita* which, with only minor exceptions, could apply to these cells in most other species of angiosperms.

Probably the most distinctive feature of companion cells is the density of their protoplasts, only a small proportion of the total volume being occupied by vacuoles. Because of the cytoplasmic density, the vacuoles tend to be sharply defined. The density of the protoplast seems largely to result from the abundance of organelles and membraneous structures. The companion cell contains a nucleus which usually disintegrates at the same time as the associated sieve element becomes moribund.

Plastids can be present, though with few internal membranes, and they have not been observed to contain starch. In some plants chloroplasts may be present in the companion cells of the aerial organs. Mitochondria are generally abundant and possess well-represented, tubular inner membranes. Dictyosomes and endoplasmic reticulum are also prominent components of companion cells. Sometimes slime is also found.

In contrast, phloem parenchyma cells, although they contain numerous cytoplasmic inclusions, organelles and a nucleus, are highly vacuolated. They are usually smaller than the sieve elements when seen in transverse sections but larger than the companion cells. Chloroplasts are often conspicuous in phloem parenchyma. The mitochondria seem to be normal with a double outer membrane and tubular inner membranes. They appear to be less numerous than in companion cells, but this may be because they are restricted to the outer layer of the highly vacuolated protoplast. The endoplasmic reticulum is very prominent in parenchyma cells.

Connections between phloem cells

Although plasmodesmata have been reported between sieve elements and both companion cells and parenchyma cells (Esau and Cheadle, 1965), it seems that the most marked connections lie between the sieve elements and companion cells. In the case of *Tilia americana*, Evert and Murmanis (1965) were unable to show plasmodesmata between parenchyma cells and sieve elements, the connection between these cells apparently being only through the companion cells. It has been suggested by these workers that the sieve element–companion cell connections may be more related to those between sieve elements than to true plasmodesmata. The structures between sieve elements and companion cells are frequently branched on the side of the companion cells (Esau and Cheadle, 1965).

Sieve elements and sieve cells

The cytological investigation of sieve elements, particularly at the electron microscope level, seems to be bedevilled by at least three profound difficulties.

Firstly, there is the problem of whether a particular sieve element under observation was functional prior to fixing of the material. We have dealt with this question in a previous section and it is not proposed to enter into further consideration of the problem, other than to urge the reader to bear it in mind when trying to assess the situation.

The other two problems which must be solved before any agreement can be attained are technical ones concerned with preparation of material. It is known that the contents of sieve elements are under a considerable hydrostatic pressure (the evidence for this is dealt with in Chapter 10); thus any procedure which involves opening the sieve tube system to the atmosphere may cause damage by inducing a surging flow in the cells, displacing the contents.

The precautions taken by different workers to avoid damage by turgor release are several. An obvious method is to reduce the turgor by osmotic solutions. Hepton, Preston and Ripley (1955) treated segments of stem with sucrose solutions, as did Mehta and Spanner (1962). Other workers have injected fixatives into intact, hollow stems (Esau and Cheadle, 1961). More recently, a number of workers (Northcote and Wooding, 1966; Evert, Murmanis and Sachs, 1966) have merely cut up the material directly into the fixative.

Almost instant freezing of intact vascular tissue might seem an excellent way to overcome the problem of turgor release. Ziegler

(1960) froze intact bundles of *Heracleum* in liquid air and then freeze dried frozen pieces prior to fixation. Johnson (1968) has also frozen pieces of intact tissue in liquid nitrogen and examined the sieve elements by use of the process of freeze etching.

Fixation of phloem tissue can also pose considerable problems. Earlier workers tended to use either chromic on osmic acids or potassium permanganate, the latter having been particularly favoured until recently by Esau and her colleagues (Esau and Cheadle, 1965). However, there is much evidence (Weatherley and Johnson, 1968; Crafts and Crisp, 1971) that these fixatives do not give critical views of plant tissues at the ultrastructural level. Permanganate, particularly, is known to cause a dissolution of certain cytoplasmic structures, frequently replacing the non-membraneous components with coarse precipitates. It now appears that glutaraldehyde and acrolein are more suitable fixatives, for structures such as plasmatic filaments can be observed when these are employed.

Clearly, the difficulties to be encountered in ultrastructural investigations are considerable. It is therefore not surprising that the range of structure described by different workers is rather broad. However, it is possible to clarify the situation to some extent, for the evidence we have available points to one of five possible configurations (*Figure 6.1*).

Slime, filaments and strands

'Slime' is a name which has been used since the earliest investigations with the optical microscope on sieve element structure. It was applied to the apparently proteinaceous material which could be found either as discrete bodies in the cytoplasm of differentiating sieve elements or as plugs at the sieve plates of mature elements.

Most workers on the ultrastructure of sieve elements now seem to agree that the lumen of mature elements contains filamentous structures, although there is no accord as to their arrangement. Some years ago, Kollmann (1960) described filaments measuring 70–130 Å in diameter in sieve elements from *Passiflora*. This worker also conclusively showed that these filaments were the same as the so-called 'slime' of the earlier optical microscopists.

The nomenclature which has been used to characterise the fibrillar system of sieve elements is still somewhat confused. Duloy, Mercer and Rathgeber (1961) used the term 'slime fibrils'; Eschrich (1963) wrote of a 'lipoprotein network'; Engleman (1965b) spoke of 'mictoplasm'; while Behnke and Dörr (1967) used the term 'plasmatic filaments' for the structures they found in sieve elements of *Dioscorea*.

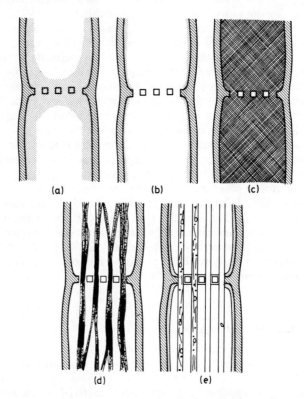

Figure 6.1 Diagrammatic representation of possible configurations of sieve element contents. (a) Pores obstructed by cytoplasm. (b) Pores unobstructed, so that empty lumina of sieve elements are connected. (c) Lumina and pores filled with a uniform network of filaments. (d) Bundles or strands of filaments lying within the lumina of sieve elements and through the pores in the sieve plates. (e) Membrane-bound transcellular strands with and without contents. (From Weatherley and Johnson, 1968, courtesy of Academic Press)

Quite recently, the term 'slime' has to a large extent been replaced by the name P(phloem)-protein, first introduced by Cronshaw and Esau (1967). These workers showed the P-protein in *Nicotiana* to arise as small groups of tubules in the cytoplasm, these subsequently enlarging to form compact masses of P1-protein consisting of tubules having a diameter of 231 ±2.5Å. During later stages of differentiation, these tubules become dispersed, finally becoming reorganised into smaller, striated fibrils, 149 ± 4.5 Å in diameter which Cronshaw and Esau designated P2-protein. In later publications Cronshaw and Esau (1968a, b) concluded that there are at least four different types of P-protein in *Cucurbita* which can undergo change from one type to another. Evert and Deshpande (1970) have demonstrated that some P-protein material can be nuclear in origin, while a number of other workers have confirmed the presence of different forms of P-protein in sieve elements (Northcote and Wooding, 1966; Behnke, 1969a).

It is clear from what has just been said that P-protein has a well-organised structure when viewed in the electron microscope. On the other hand, the term 'slime' is suggestive of an amorphous material. Thus the question arises as to whether P-protein is the same as 'slime'. As mentioned previously, 'slime', historically, was a name given to material which could be seen in sieve elements with the optical microscope, and, since it appeared to be structureless, it was generally considered to be a product of the breakdown of sieve element protoplasts which occurred during maturation of these cells. That is, as far as solute movement was concerned, it was thought to be inert.

It may well be that some of the 'slime' seen in the sieve elements of certain species with the optical microscope is indeed an inert substance. Walker and Thaine (1971) showed that exudate from *Cucurbita* sieve tubes (the same genus as used by Cronshaw and Esau, 1968a and b, in their work on P-protein) contained, not only a structural fraction composed of fibrillar protein, but also a fraction which gelled when exposed to the atmosphere. The resulting solid from this gelling process appeared amorphous in the optical microscope and even the electron microscope failed to reveal any organised fine structure. 'Slime', then, may consist of several components, at least in some species: an inert fraction composed of coagulated protein and any products produced by degradation of the sieve element protoplasts, and a P-protein fraction composed of organised protein filaments and tubules. Since the latter component clearly has a well-defined sequence of events in its formation and its final structure appears highly organised, it would seem, at any rate on a subjective basis, that P-protein, unlike the inert components of 'slime', might have a fundamental role to play in the transport process.

Our understanding of the role of P-protein is, however, not helped by the fact that there is as yet no accord as to the arrangement of this substance within the lumen of the sieve element. Esau and Cheadle (1965) have pointed out the difficulties involved in trying to relate the form of structures seen in the electron microscope in fixed and sectioned material to the form which the structures take in living protoplasts. As a rather broad generalisation, it can be said that most workers favour the concept of a reticulum of P-protein fibrils occupying the lumen of sieve elements, in the manner represented diagrammatically in *Figure 6.1 (c)*.

There are a number of reports in the literature which suggest that protein filaments may be aggregated into bundles or even discrete strands. Kollmann, Dörr and Kleinig (1970) have published electronmicrographs of the exudate from *Cucurbita* phloem showing 90 Å filaments grouped together to form broader fibrils. However, it is possible, as Kollmann and his colleagues point out, that aggregation of fibrils may have occurred as a result of the conditions of their experiments.

Working with the optical microscope, Parker (1964) and Evert and Derr (1964) have demonstrated strands which traverse the lumina and sieve plates of the sieve elements of a number of tree species. Evert and Murmanis (1965) have reported finding strands in sieve elements of *Tilia americana* in studies with the electron microscope.

The most vigorous proponent of discrete strands is Thaine. In 1961 this worker, using the optical microscope to observe *living* phloem of *Primula obconica*, described structures termed 'transcellular strands' which traversed the lumina and sieve plates. He described particles which moved along these strands and observed the particles to move through a file of at least 10 elements. In a later paper Thaine (1962) reported the strands to be between 1 and 7 μm diameter in *Cucurbita* and compared them to transvacuolar strands of cytoplasm found in other cells. He also suggested that they were bounded by a surface membrane and contained mitochondria-like particles and small plastids (*Figure 6.1 e*). *Figure 6.2* presents a series of photographs taken by Thaine, Probine and Dyer (1967) of these transcellular strands.

Thaine's observations were strongly criticised by Esau, Engleman and Bisalputra (1963), who claimed that he had merely seen diffraction lines from cell walls, i.e. that strands do not exist. However, Thaine and his colleagues (1967) have effectively countered this argument, for they were able to demonstrate that the strands came into and out of focus as the objective was focused up and down through a sieve element (*Figure 6.2*). Moreover, strands may be seen with the interference microscope; thus they presumably must be real structures.

Certainly, there is a good deal of evidence in favour of cytoplasmic material in the form of strands and Thaine's conclusions are undoubtedly made more cogent by the fact that his observations were

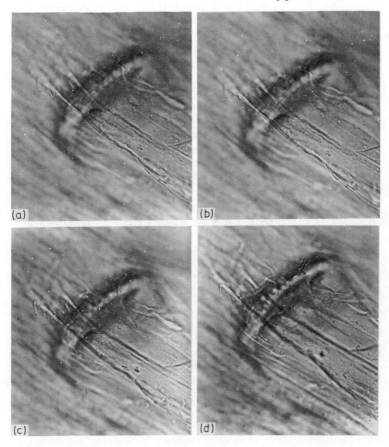

Figure 6.2 A through-focus series of 2 μm *steps through the sieve plate of* Cucurbita pepo, *showing strands which come in and out of focus as the focusing depth is deepened from* (a) *to* (d). *(From Thaine, Probine and Dyer, 1967; reproduced by courtesy of the authors and The Clarendon Press)*

made on living sieve elements. In a recent paper Jarvis and Thaine (1971) argued that strands might be more easily observed in dead elements if these were frozen and then sectioned, rather than chemically fixed. Pieces of *Cucurbita* phloem were frozen in liquid nitrogen, sections then being cut on a cryostat. When the sections were subsequently thawed, strands 5–9 μm wide could be seen with the optical microscope. According to Jarvis and Thaine, these strands showed a discrete boundary enclosing groups of parallel structures and were only preserved when freezing was rapid.

What appears to be rather disturbing about Thaine's ideas, however, is the lack of corroborative evidence from electron miscroscopy. Tamulevich and Evert (1966) studied the sieve elements of *Primula obconica*. Although they believe that strands exist in this species, they were unable to find evidence for the form of these as visualised by Thaine. Of course, it is very possible that Thaine's strands could be destroyed by the techniques used in sample preparation for the electron microscope, disintegrating to form a network of P-protein filaments. Johnson (1968) attempted to overcome this problem by rapid freezing of phloem tissues which were still attached to petioles of *Nymphoides*, followed by examination of the sieve elements by a freeze-etching technique. However, he could find no evidence for membrane-bound strands.

The evidence for discrete, rather complex transcellular strands as visualised by Thaine remains equivocal. Until more evidence is obtained, it is not possible to be certain that the strands observed by many investigators are at all comparable to those described by Thaine.

Sieve plates

The most fundamental question we have to answer about the sieve plates concerns the state of the sieve pores: Are these 'open', giving complete continuity from one sieve element lumen to the next, or are they occluded to a varying degree by fibrillar structures or callose? Attempts to answer this question have given rise to some of the most heated arguments, based largely on questions of technique, between workers in the ultrastructural field. What does seem fairly certain is that the sieve pores are not closed by a membrane, i.e. structure (*a*) of *Figure 6.1* would be discounted by virtually all present-day cytologists.

Esau and Crafts have always been firm supporters of the 'open pore' concept, the arguments in favour of this having been given by Crafts and Crisp (1971). As these authors point out, many of the earlier workers used permanganate fixation, and this may have led to the conclusion that the sieve pores were filled with what appeared to be solid material (Hepton and Preston, 1960; Kollmann, 1960). Nonetheless, Esau and Cheadle (1961) have published electron-micrographs of sieve pores in *Cucurbita* (Figures 9 and 10 of their paper), showing these to be apparently open in material fixed with permanganate.

There is no doubt that new fixation techniques have shown in a number of instances examples of sieve plates with apparently open pores, although Bouck and Cronshaw (1965) found dense material plugging the sieve pores of *Pisum* in material fixed with glutaraldehyde or acrolein. In a later paper Cronshaw and Anderson (1969) investigated the effects of different fixation procedures on the state of the sieve pores in *Nicotiana*. Material fixed with glutaraldehyde or glutaraldehyde–formaldehyde at room temperature showed sieve plate pores to be filled with P-protein. Acrolein-fixed material gave a similar picture. In contrast, in material which had been frozen rapidly and then chemically fixed at low temperature, many sieve plates with 'open' pores or only loosely filled with P-protein were observed (*Figure 6.3*). In other publications (Cronshaw, 1969; Anderson and Cronshaw, 1970) illustrations are given of open pores.

In contrast, it is possible to find many illustrations, also given in recent papers, of occluded or partially occluded sieve pores. Johnson (1968) and Weatherley and Johnson (1968) have shown electromicrographs of sieve pores containing filaments in which the material was either frozen and freeze-etched or fixed in glutaraldehyde. Mishra and Spanner (1970) and Siddiqui and Spanner (1970) argue that sieve pores are largely filled with P-protein. Most recently, Evert *et al.* (1971) have indicated that the sieve pores in sieve elements of leaf veins of *Hordeum* are filled with endoplasmic reticulum. Behnke (1971), following a cold fixation technique with glutaraldehyde–acrolein, has produced some excellent electronmicrographs of the sieve plates in *Aristolochia* in which both protein filaments and endoplasmic reticulum extend through the pores (*Figure 6.4*).

What does appear certain about this confusing situation is that much of the argument about 'open' or 'blocked' pores is a matter of definition, for the various workers in the field seem to differ considerably over what is meant by these terms. 'Open' pores almost invariably show filaments in the pores, but the argument seems to be that spaces occur between these filaments which ensure continuity of the contents of two contiguous sieve elements. Conversely, 'blocked' pores also usually show interfilamentar spaces. How one defines these terms depends to a large extent upon the hypothesis of the mechanism of movement one supports. Presumably, supporters of the pressure flow mechanism either disregard the loose filaments in 'open' pores or believe that they are artefacts caused by fixation procedures or that the filaments are themselves mobile. Opponents of pressure flow, although their chosen hypothesis may require spaces between filaments, believe that these are too small to allow a pressure flow to occur. Thus it is quite possible for two different workers to look at the same sieve pore, and for one to say it is 'open' and the other that it is 'blocked'!

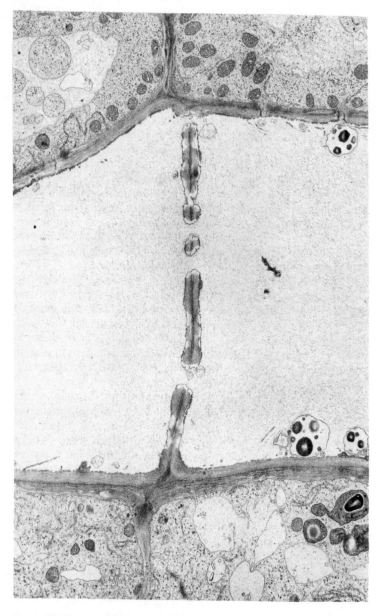

Figure 6.3 Electronmicrograph of a sieve plate from a wilted plant of Nicotiana showing open pores. (Cronshaw, 1974; reproduced by courtesy of the author and McGraw-Hill Book Company (UK) Ltd)

The situation becomes even more complicated when arguments about the extent of callose deposition on sieve plates are considered. If filaments are present in 'open' pores, then a heavy deposition of callose will make the pore appear more 'blocked'. Eschrich (1970) has pointed out that a close relationship must exist between the density of pore filaments and the extent of the callose cylinder. Killing whole plants by freezing (Shih and Currier, 1969) or by *in situ* fixation (Eschrich, 1963) prevents the occurrence of heavy callose deposits. The pores therefore appear more 'open', since the filaments are not then bundled together. Cronshaw and Anderson (1969) have also demonstrated that the extent of callose deposits depends upon fixation procedures. It may well be that callose is not normally present in uninjured sieve elements; therefore the pores are more 'open' than those observed in fixed material.

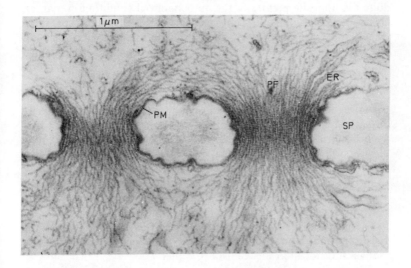

Figure 6.4 The sieve pore contents in Aristolochia brasiliensis (*Mart. et Zucc.*). *Material cold-fixed in glutaraldehyde–acrolein. ER, endoplasmic reticulum; PF, plasmatic filaments; PM, plasma membrane; SP, sieve plate. (From Behnke, 1971; reproduced by courtesy of the author and Academic Press)*

Clearly, it is possible to argue about the situation *ad infinitum* in the present state of our knowledge. We must now leave the subject of the sieve pores, returning to them in the following chapter, where the possible structures will be examined in the light of the various hypotheses.

Mitochondria and plastids

A certain amount of controversy continues with regard to the condition of the mitochondria in mature sieve elements. Esau and Cheadle (1962a) have reported that while the mitochondria of young, nucleate sieve elements of *Cucurbita* have the usual mitochrondrial structure, those of mature sieve elements appear to be somewhat modified. During sieve element differentiation, these organelles show a disorganisation of the inner membranes and possibly also of the outer membrane. Esau and Cheadle came to the conclusion that the mitochondria of mature elements are 'more or less degenerate'. This view has been supported by the results of other workers (Engleman, 1965b; Evert and Murmanis, 1965).

On the other hand, there are reports in the literature which give quite the opposite picture of the mitochondrial condition in mature elements. Kollmann and Schumacher (1964), Behnke (1965) and Bouck and Cronshaw (1965) have all come to the conclusion that mitochondria show completely normal appearance in functioning sieve elements. Evert *et al.* (1966) have demonstrated normal mitochondria in *Cucurbita*.

These differences of interpretation may be caused, like many others in the field of phloem ultrastructure, by the fixation techniques employed. Such a possibility has been raised with regard to mitochondria by Tamulevich and Evert (1966). However, Esau and Cronshaw (1968a) have pointed out that if certain abnormalities they observed in sieve element mitochondria were indeed artefacts of fixation, then this must mean that these organelles are also more sensitive to manipulation than those of other cells which appear perfectly normal.

The problem of the mitochondrial condition needs to be solved. It is important from the point of view of the energy supply to any 'active' translocation mechanism (Chapter 7).

The question of the occurrence of plastids in functioning sieve elements is, fortunately, not one which has aroused a great deal of controversy; it seems now to be generally agreed that these structures are present, anchored to the parietal layer. In 1964 Kollmann concluded that plastids were to be found in mature elements of many angiosperms (*Cucurbita* being an exception) and gymnosperms. Plastids have now, however, been found in the sieve elements of *Cucurbita* (Esau and Cronshaw, 1968a).

Starch appears to be present in many cases (Behnke, 1969b; Evert and Deshpande, 1971) but absent in others (Esau and Cronshaw, 1968a).

The tonoplast (vacuolar membrane)

It was realised some time ago that the presence or absence of a tonoplast in functioning sieve elements could markedly affect our views on the mechanism of transport; if the sieve pores were closed by a membrane, then it seems unlikely that any hypothesis requiring a bulk flow of solution would be viable. Esau and Cheadle (1962b) could find no evidence for the existence of a tonoplast in mature sieve elements of *Cucurbita*. They discounted the possibility that their findings were caused by fixation techniques, for they were readily able to discern a tonoplast in other phloem cells. Of course, this latter conclusion may not be a valid one; if the internal conditions such as pressure, pH and the chemical environment in sieve elements were quite different from other phloem cells, as they may well be, then fixatives could have a different effect on sieve elements than on other cells.

In fact, however, most workers seem to agree that the tonoplast breaks down at about the same time as the nucleus disintegrates, although Kollmann (1960) has claimed to have found a tonoplast in mature, enucleate sieve elements of *Passiflora*. Tamulevich and Evert (1966) and Evert *et al*. (1966) have found what they termed a 'delimiting membrane' separating the parietal layer of cytoplasm from the cavity in *Primula* and *Cucurbita*, respectively. There does not appear to be any suggestion that the 'delimiting membrane' occludes the sieve pores, for Tamulevich and Evert (1966) believe that this membrane lines the pores, together with the plasma membrane.

Assuming that the tonoplast disintegrates, it does not seem possible to define the central region of a mature sieve element as a vacuole, and the term 'lumen' is generally applied. Following on from this, there has been considerable discussion as to whether the contents of the lumen are cytoplasmic or vacuolar. Nothing of substance has emerged from this, except a number of terms such as 'dilute cytoplasm' and 'mictoplasm', which have only served to underline our ignorance of sieve element structure.

Endoplasmic reticulum

As with the case of mitochondria, there remains a continuing controversy over the condition of the endoplasmic reticulum in mature sieve elements, some investigators claiming the presence of endoplasmic reticulum of normal appearance, while others believe that this component breaks down into numerous vesicles. Once again, it may well be that the degenerate state of the endoplasmic reticulum may be more apparent than real, caused largely by variations in techniques for the preparation of material.

It could be significant that workers such as Esau have recently (Esau and Cronshaw, 1968b) approached the views of Kollmann (1964) in indicating that mature sieve elements can possess a reasonably well-developed endoplasmic reticulum. Esau and Cronshaw (1968b) describe the endoplasmic reticulum as consisting of two forms, one a parietal network closely applied to the plasma membrane, the other consisting of stacked membranes. Certainly, a number of investigators have claimed that the endoplasmic reticulum can appear as a well-organised structure in mature elements (Tamulevich and Evert, 1966; Behnke, 1971).

Phloem plugging

It has been mentioned in several of the preceding chapters that exudation of a solution, rich in sugars and other solutes, can be obtained from the phloem of many species of plants, by either making gross incisions into the phloem or using aphid stylets to tap individual sieve elements. Although exudation from stylets can proceed for considerable periods of time (5 days appears to be the record for *Tuberolachnus* stylets on willow), the production of liquid from phloem incisions is generally a short-lived phenomenon, lasting only for a few hours at the most.

The most plausible explanation for the cessation of exudation from incisions is that the injured sieve elements become plugged by some material. This possibility is supported by the observation that exudation (for instance, from cucurbit stems) may be restarted by taking a thin slice of tissue from the edge of the original incision. Moreover, phloem exudation cannot be obtained from many species; thus it can be assumed that the plugging mechanism must be very efficient in these cases.

If sieve elements are capable of being rapidly and effectively sealed, it may be asked why exudation from aphid stylets is able to proceed for such long periods. It is possible to think of several feasible explanations. One possibility might be that the plant species which provide continued stylet exudation do not have a particularly effective sealing mechanism. A second possibility is that the aphids introduce materials into the stem (for instance, certain hormonal compounds) which stop the sealing mechanism functioning. This does not imply that aphids are able to pump materials directly into the sieve elements. Indeed, this would be out of the question, considering the high positive pressures which must be present in these cells (Chapter 10). But it would seem possible for aphids to secrete substances in the vicinity of the sieve elements, after which these materials could move into the conducting cells.

It is certainly possible to envisage the plugging system(s) as being under hormonal control. Vigorous exudation can be obtained from *Yucca* flower stalks, and will proceed for several weeks if the incision is renewed twice a day (Tammes and van Die, 1964). Here it could be that changes in the level of naturally occurring plant hormones which are responsible for flowering may also be affecting the plugging mechanism. The induction of exudation from incisions into stems of *Ricinus* by massage pretreatment (Milburn, 1970) could well be mediated through hormonal agencies stimulated by the massage. Hormonal control of plugging mechanisms in *Ricinus* appears more probable from the observations of Hall, Baker and Milburn (1971) that exudation can be obtained without massage pretreatment if vigorously growing plants are used.

To understand the third reason why stylet exudation can be so prolonged, it is necessary to enquire how the plugging mechanisms are triggered into operation. Seemingly the most plausible explanation would be to assume that the sudden fall in hydrostatic pressure, consequent upon opening the sieve elements to the atmosphere, could initiate some form of sealing. Therefore, if the drop in hydrostatic pressure were small, the sealing processes might not be stimulated. The small diameter of the food canal in aphid stylets (2 μm in *Tuberolachnus*), giving a high resistance to flow, would thus produce only a small decrease in pressure in the pierced sieve element, a situation which would not occur with gross incisions.

The fact that high positive pressures exist in sieve elements has led to a rather teleological explanation as to why sealing mechanisms have been evolved. It is argued that if the sieve elements could not be plugged, plants would 'bleed to death' after extensive injury by, for instance, grazing animals.

Plugging mechanism

Judging from the variation in the time which is taken for sieve element sealing in different species, it appears very probable that several mechanisms may be involved, some being extremely rapid, e.g. in species which cannot be induced to exude, others more protracted in their response.

Rapid plugging could occur as a result of collapse of the sieve plate consequent upon a drop in hydrostatic pressure after incision of the sieve elements, leading to a considerable reduction in the diameter of the sieve plate pores. If the lumen and sieve pores are filled with a reticulum of fibrillar structures, then it can readily be envisaged that these could be 'blown' into the constricted sieve pores, thus rapidly and effectively sealing the cut.

A somewhat slower plugging mechanism could be callose formation on the sieve plates, causing complete blockage of the sieve pores. It is not yet known whether callose is a normal component of the sieve plates of functioning sieve elements. As mentioned before, it seems that the amount of callose deposits depends upon the speed at which phloem tissues are killed. The killing of whole plants by rapid freezing (Shih and Currier, 1969) leads to very small callose deposits, while no callose at all could be detected with other *in situ* fixation techniques (Eschrich, 1963; Evert and Derr, 1964). Further evidence that injury to the phloem can result in heavy deposition of callose comes from the work of Engleman (1965a), who showed the sieve elements of *Impatiens* to be affected by callose formation for a distance as great as 15 cm from an incision.

Some controversy exists as to the effect of callose deposition on the extent of transport. Eschrich, Yamaguchi and McNairn (1965) induced callose formation in cucurbit plants by treatment with calcium chloride and boric acid solutions. However, they could not detect any effect of the callose upon the transport of ^{14}C-labelled assimilates, or of fluorescein. It seems possible that insufficient of the sieve tubes in their experiments were closed by callose to produce a noticeable effect upon transport. Callose formation can also be induced by heat treatment. Currier, McNairn and Webster (1966) demonstrated an effect of heating cotton hypocotyls to 45°C for 15 min on callose formation and transport of labelled assimilates; callose increased and transport decreased. McNairn and Currier (1968) showed that transport of ^{14}C-assimilates decreased immediately after heat treatment, with a recovery after several hours. This was correlated with an increase and subsequent decrease in the amount of callose on the sieve plates.

It seems most probable that heavy callose deposits on the sieve plates can effectively block translocation, and that callose formation constitutes a quite rapid and effective method for sealing injured sieve tubes.

A third mechanism of sieve element plugging is the distinct possibility that the contents of sieve elements of certain species coagulate under aerobic conditions. Northcote and Wooding (1966) envisaged a sol–gel conversion. Phloem exudate from *Cucurbita* rapidly becomes turbid and coagulates (Crafts, 1932). Milburn (1971) demonstrated an aerobic coagulation of phloem sap from *Cucurbita*;

exudate sealed in capillary tubes remained clear and liquid for many weeks, but exposure to the air brought about rapid coagulation. Walker and Thaine (1971) found that *Cucurbita* exudate did not gel when it was collected in a buffer containing 2-mercaptoethanol or dithiothreitol, both of which are —SH group reducing agents. They came to the conclusion, therefore, that coagulation was due to the formation of S—S between soluble proteins under oxidising conditions. The gel thus formed had no organised fine structure when viewed with the electron microscope.

It is easy to envisage that a combination of all three mechanisms —collapse, callose formation and protein coagulation—could rapidly and effectively seal injured sieve tubes. Recovery from extensive injury to secondary tissues could be brought about by the formation of new phloem.

7

The hypotheses of the mechanism of sieve tube transport

Over the past 100 years or so, a number of mechanisms have been suggested to account for the phenomenon of phloem transport, the three most recent having been initiated during the last 15 years. The efforts of those physiologists who have elaborated hypotheses should certainly not be decried, but it is a sad reflection upon the state of our knowledge that it is far easier to propose a mechanism than to prove or disprove its existence. Indeed, of the various proposals which have been made, it seems that only one, or possibly two, are now generally discounted.

The purpose of the present chapter is to describe the characteristics of the various hypotheses in relation both to the ultrastructural picture, which has already been reviewed, and to the physiological data, which will be dealt with in subsequent chapters. Unfortunately, several of the hypotheses are rather vague on certain important points; thus difficulties arise in assessing the quantitative aspects of the proposed mechanism. Nonetheless, it is possible with certain of the hypotheses to state within fairly narrow limits what the ultrastructure of the conducting cells must be in order for the mechanism to function, and also to predict the physiological characteristics of the mechanism.

Whatever mechanism is proposed to explain phloem translocation, energy will have to be supplied to the system at some stage. On the basis of the site of energy application, it is possible to assign any hypothesis to one of two groups, a 'passive' or an 'active' group. Before proceeding to describe the hypotheses in detail, we must define exactly what is meant by the terms 'active' and 'passive', for many of the experiments designed to discriminate between the hypotheses have attempted to discover where energy is applied to the transport mechanism.

Very broadly, a 'passive' mechanism may be defined as one in which it is not necessary to apply energy directly to the solute molecules, *while they are undergoing tranport within the sieve elements.* In the passive hypotheses the energy to drive the translocation mechanism is expended at either end of the system in the pumping in of solutes at a source, or their pumping out at a sink (*Figure 7.1*). This does not, of course, imply, as many investigators seem to have believed, that the sieve tube merely acts as an inert pipe which requires no energy in order to support a movement of materials. Since the sieve tube contains complex structures, it is almost certain that some energy is required for their proper functioning, e.g. the plasma membrane would very possibly change in permeability, which would lead to leakage of solutes, if the energy necessary to maintain its structural integrity were not available.

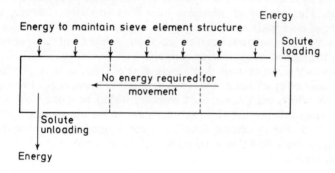

Figure 7.1 Diagram showing sites of energy application to passive transport system

In several aspects of their energy requirements the active hypotheses are similar to the passive types. Energy will be required at either end of the system to pump solutes into or out of the conduits, and energy will certainly have to be expended in order to maintain the structure of the sieve elements. However, in the active hypotheses additional amounts of metabolic energy will also have to be applied directly to the solute molecules in transit, and it is this feature which separates the two types so distinctly.

A second way in which hypothetical mechanisms may be classified is in terms of whether or not a bulk flow of solution is envisaged. The reason for raising this point at this juncture is to enable us to look very briefly at the definition of a process which has been given the somewhat formidable name of 'simultaneous bidirectional movement'. Although there is nothing intrinsically complicated about the concept of simultaneous bidirectional movement, the exact meaning of the term has led to some confusion of thought in some workers not well-versed in phloem physiology.

Figure 7.2 presents three diagrams, each showing a different route by which two solutes can move simultaneously in opposite directions in the phloem. In the case of diagram (*a*) we have a true instance of simultaneous bidirectional movement. Two different solutes are pictured as moving in opposite directions *within the same sieve tube*, each having its own source and sink. *Figure 7.2(b)* shows the situation in which two solutes, although moving in opposite directions within the same sieve tube, move towards a common sink, and therefore do not pass one another as do the solutes in case (*a*). Diagram (*c*) presents a second situation which could lead to the conclusion that 'true' simultaneous bidirectional movement was taking place. Here we have two solutes moving simultaneously in opposite directions, but in separate, though possibly, adjacent, sieve tubes. The tapping of one sieve tube by a sink (for instance, the stylet of an aphid) would lead to the appearance of both solutes at the sink owing to lateral movement of one solute from its sieve tube into the tapped neighbouring conduit.

It would seem obvious that a bulk flow of solution must preclude any possibility of simultaneous bidirectional movement. The only way in which both could exist together would be if the flow were confined to the lumen while another mechanism transported certain solutes in the cytoplasm (this has been suggested), or if several separate bulk flows were to occur within separate channels in the same sieve tube.

'Passive' mechanisms

Thermal diffusion

Movement of solutes by simple thermal diffusion can hardly be deemed to constitute a hypothesis, since it has been acknowledged for many years that this process is far too slow to account for the rates at which solutes move through the phloem. Nevertheless, it is instructive briefly to consider diffusion, since it can be readily examined in a quantitative manner and shown to be quite inadequate.

Mason and Maskell (1928b) in one of their classic researches on cotton came to the conclusion that although phloem transport showed many similarities to a diffusive process, the rate of movement in phloem was far greater than could be accounted for by thermal diffusion. Indeed, on the basis of the sugar concentration gradients which they found, Mason and Maskell were able to calculate that phloem transport was some 20 to 40 thousand times faster than could occur because of sucrose diffusing in water, being almost identical with the diffusion constant for molecules the size of sucrose diffusing in air!

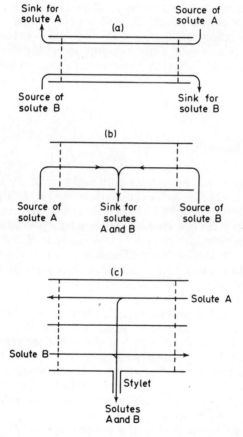

Figure 7.2 Schematic representation of possible routes of two solutes moving in opposite directions in the phloem. (a) 'True' simultaneous bidirectional movement. (b) Two solutes moving towards a common sink in the same sieve tube. (c) Two adjacent sieve tubes transporting in opposite directions with lateral movement of a solute from one sieve tube to the other. (From Peel, 1972a, courtesy of Academic Press)

Not only is diffusion along the whole length of the sieve tube quite inadequate to maintain the observed rates, it is also impossible to envisage the situation in which diffusion could account for transport across the sieve plates alone. Weatherley and Johnson (1968) have calculated that, assuming the thickness of a sieve plate to be 1 μm and taking the diffusion coefficient of sucrose in water as 0.5×10^{-5} mol cm^{-2} s^{-1} mol ΔC^{-1}, a gradient of about 6% in sucrose concentration would be required across each sieve plate. With 500 or more sieve plates per centimetre of sieve tube, this would represent an enormous over-all gradient, which could not possibly be developed. In fact, from the data of Mason and Maskell (1928b), the actual gradient at each sieve plate would be about 3×10^{-3}%. This would allow transport across the sieve plates to proceed only if the latter structures were around 5 Å thick, a most unlikely situation!

Interfacial movement

The possibility that solute transport might occur by means of a rapid movement of molecules at interfaces within the sieve element was really first suggested by van den Honert (1932). His suggestion was based upon the observation that when a drop of water-insoluble liquid whose molecules contain a polar group is placed on a water surface, the substance will spread rapidly over the surface. Van den Honert, in fact, carried out experiments with a model system consisting of two glass bulbs connected by a tube to demonstrate the velocity at which movement can occur. A water/ether interface was made within the apparatus by half filling it with acidified water containing an indicator, on to which was layered ether. Alkaline potassium oleate was then added to one end of the system, the rate of spread of this substance being measured by observation of a change in the colour of the indicator. High velocities were found (greater than 120 cm/h) at least over the first part of the pathway, although much lower velocities were measured as the distance from the 'source' was increased.

Over the years, the interfacial movement hypothesis has received very little support, the majority of investigators having discounted surface migration phenomena for a variety of reasons. Mainly, however, doubts have been raised concerning the capability of interfacial movement to account for the rates of mass transfer known to occur in phloem, and also as to whether a sufficient variety of different interfaces could possibly be present within the sieve element to transport the extremely wide range of chemical species which are phloem-mobile.

During recent years, several workers have raised the possibility that some transport, albeit a rather small proportion of the total, may be mediated via surface movement. Nelson *et al.* (1958) claimed to have observed a velocity of transport as high as 7 m/h after supplying $^{14}CO_2$ to the primary leaves of young soyabean plants, radioactivity being detected in the roots only a few seconds after the application of the tracer. In a later publication Nelson (1962) suggested that these very high velocities might have been brought about by interfacial movement of the labelled assimilates, while a second, much slower component of the translocation system, operating at velocities around 1 m/h, could be mediated through another mechanism.

Fensom (1972) has discussed the results of experiments in which tracers were injected into the phloem strands of *Heracleum*, using a technique developed by Fensom and Davidson (1970). Although Fensom believes that the bulk of the tracer movement is mediated by the contractile action of lipoprotein strands, to which we shall refer later, he suggests that the experimental results also indicate the participation of a small, surface-layer component of translocation, operating at velocities above 1000 cm/h.

Lee (1972) has been moved to apply a theoretical treatment to the surface movement hypothesis by the numerous observations of protein filaments in sieve elements. Assuming the filaments to occupy 10% of the cross-sectional area of the sieve element lumen and to have a mean diameter of 120 Å, and taking the dimensions of a sieve element to be 250 μm × 20 μm, Lee calculates the effective surface area of the filaments in one element to be 2.65×10^{-2} cm². If sucrose is present within the element as a 10% solution, there would be 1.41×10^{13} molecules of sucrose per element. Using a value of 5.3Å for the radius of a hydrated sucrose molecule, Lee further calculates that the surface area of the filaments would support a monolayer of 2.36×10^{12} molecules, i.e. within an order of magnitude of the total number of sucrose molecules in the sieve element.

Whatever else these calculations show, they at least throw doubt upon the statements of those workers who have discounted the interfacial movement hypothesis on the grounds of lack of sufficient surface area. Also, as Lee points out, surface movement phenomena could explain simultaneous bidirectional movement within a single sieve tube, and could account for minimal water movement if only the hydration spheres of sucrose molecules were moved.

Pressure flow

There is no doubt that the pressure flow hypothesis, the modern formulation of which was given by Münch in 1930, has been the most widely accepted of the many and varied proposals which have been put forward. The main reason for the popularity of the pressure flow concept is, without question, its inherent simplicity. This has rendered it acceptable, not only to many workers within the field of phloem physiology but also to a considerable number in other areas of botanical research. Although the simplicity of the pressure flow mechanism allows it to be easily understood and readily quantified, it has also worked against the universal acceptance of the hypothesis; many investigators doubt the operation of such an uncomplicated mechanism in the apparently complex system of the phloem.

The fundamentals of the pressure flow hypothesis can be readily understood by reference to the now famous diagram of a model system shown in *Figure 7.3(a)* (Münch, 1930). Here we have two vessels A and B (for instance, porous pots), the surface of each vessel being covered by a semipermeable membrane and connected by a glass tube. Vessels A and B are surrounded by water, and initially both vessels and their connecting tube are filled with water. If by some means a solute such as sucrose is now introduced into vessel A, this will cause a lowering of the diffusion potential of the water, and water will move into vessel A by osmosis from the external milieu, this leading to an increase in hydrostatic pressure in A. Consequent upon this pressure increase, water will be forced along the connecting tube to B, carrying sucrose with it. The subsequent rise in hydrostatic pressure in B will in turn cause a diffusional movement of water out of vessel B. These processes would normally continue until the sucrose concentrations in A and B were the same, the system then reaching equilibrium. If, however, we were able continuously to apply energy to the model by adding sucrose to A and removing it at B, we would produce a continual transport of water and sucrose from A to B. It should be noted that no energy need be continuously applied to the transport channel (the glass tube).

Clearly, such a model, given the correct conditions, will function. Moreover, the principles embodied in this model can readily be applied to the plant (*Figure 7.3b*). At one end of the system is a source—for instance, an assimilating leaf cell—which is secreting solutes into one end of the sieve tube, water movement following osmotically from the xylem elements. The turgor pressure thus generated drives the solution through the sieve tube to the sink cell, where the solutes are removed. The water moving out of the sieve tube at the sink end is channelled back into the xylem and is thence transported back to the leaves.

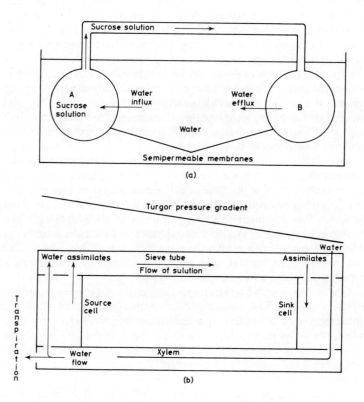

Figure 7.3 (a) Model illustrating the working of the pressure flow mechanism. (b) The model as applied to the vascular systems of higher plants

Thus it is possible to 'see' how pressure flow could work, and also to state quite categorically what the physiological characteristics of phloem transport must be if it is brought about by this mechanism. These characteristics are as follows.

Pressure flow is a typical passive mechanism as passive has been defined, the major sites of energy application being at either end of the system, although there seems little doubt that some energy must be supplied throughout the length of the sieve tube to maintain the semipermeability of the sieve elements. For pressure flow to operate, there must be a gradient of osmotic potential along the sieve tube in the direction source→sink, and this must be capable of generating a turgor gradient in the same direction.

Unlike the other two passive mechanisms (thermal diffusion and interfacial flow), pressure flow involves a bulk movement of solution with both water and solutes moving together. This fact has very important physiological considerations, for it means that any solute which enters the translocation stream must move, not necessarily along its own concentration gradient, but along the total concentration gradient of all the solutes present. Thus it should be well-nigh impossible for a solute to enter a mature exporting leaf against the stream of assimilates which is pouring out to the sink regions of the plant. In other words, pressure flow would preclude any possibility of simultaneous bidirectional movement taking place in a single sieve tube.

It also follows that the molecules of all the solutes present within a sieve tube must move with the same velocity, unless certain species are impeded, by, for instance, electrostatic effects in charged sieve pores. It must be emphasised here that it is velocity, not mass transfer, to which this statement refers. It is quite possible with the pressure flow mechanism for two species to move with identical velocities while they are actually within the sieve tube, but to have two different over-all mass transfer rates caused by differences in their rate of removal from the conduits (*Figure 4.4*, p. 75).

Just as it is possible to enumerate the physiological characteristics of the pressure flow mechanism, it is also possible to state with some degree of certainty what the ultrastructure of the sieve tubes would have to be in order to render the hypothesis energetically feasible. To put the situation in a nutshell: if the sieve pores are open, then pressure flow is possible; if the pores are occluded by filaments, then it is out of the question.

It seems almost certain (Horwitz, 1958) that flow in the sieve tubes would be laminar rather than turbulent; thus the Poiseuille equation,

$$P \times \frac{8 \eta \, V l}{r^2}$$

where P = pressure difference (dyn/cm^2) between the ends of a tube of length l (cm) and radius r (cm), through which a solution of viscosity η (P) is flowing with a velocity of V (cm/s), can probably be safely applied to the system.

Crafts and Crisp (1971) have applied the Poiseuille formula in an attempt to calculate the pressure gradient required to drive a 10 % sucrose solution at a velocity of 130 cm/h through the sieve tubes of *Cucurbita melopepo*. Taking the radius of the sieve pores as 2.41 μm and the thickness of the sieve plate as 5 μm, and assuming these to be open (although they discount the area said to be occupied by *plasmatic filaments*), they arrive at a value for the pressure gradient required of 92×10^{-6} atm per sieve plate, or 0.37 atm/m when the effect of the lumina of the sieve elements is taken into account.

According to Crafts and Crisp, this figure may be too high by a factor of as much as 2 to account for the measured velocities, for, as they point out, the openings of the sieve pores are not angular, but curving. This, they argue, must mean that flow, instead of being laminar for the whole length of the pore, converges at the entrance and diverges at the exit; therefore the Poiseuille formula cannot be strictly applied, and they derive an equation which gives a value only half that derived from the Poiseuille formulation.

Crafts and Crisp also draw attention to the situation in which the diameters of the sieve pores are equal to or greater than the sieve plate thickness. Under these conditions, flow may be creeping rather than laminar. The formula for creeping flow cited by Crafts and Crisp from a publication of Happel and Brenner (1965) is

$$P = \frac{3\pi\, V \eta}{r}$$

which produces a lower value for the required pressure gradient than the Poiseuille expression. Crafts and Crisp believe that the above formula is particularly applicable to the situation in the phloem of leaves and fine roots, although in the trunks of trees the Poiseuille formula is probably more appropriate.

Calculations by other workers in which the sieve plate pores are assumed to be open give values for the required pressure gradient of the same order of magnitude as those derived by Crafts and Crisp (1971). Weatherley and Johnson (1968) reach a figure of 0.6 atm/m for flow through sieve tubes of willow of a 10% sucrose solution at a velocity of 100 cm/h. Since the osmotic potential of the sieve tube exudate from willow is around 15 atm (Peel and Weatherley, 1959), this would enable pressure flow to work satisfactorily in trees up to 25 m in height. However, in very tall trees either the osmotic potential of the sap would have to be very high or the velocities of transport lower than 100 cm/h; thus Weatherley and Johnson conclude that in these extreme cases the operation of a pressure flow mechanism is doubtful. Crafts and Crisp (1971) have countered this argument by pointing out that very tall trees do have high sap concentrations, and also that the roots of these trees are probably supplied by the lowest branches; movement, therefore, does not need to occur over the whole length of the trunk.

While there seems little doubt that pressure flow is feasible if the sieve pores are completely open, i.e. with no filaments or other cellular components passing through them, the situation would appear to become quite different if pores partially occluded by fibrils are considered. Weatherley and Johnson (1968) quote a theoretical approach to this situation formulated by Spanner (published in Fensom and Spanner, 1969), in which fibrillar filled pores are considered in terms of an hexagonal distribution of parallel rods. Taking the filaments as being 100 Å thick and lying 200 Å apart, application of Spanner's formula to sieve plates 1 μm thick gives a value for the pressure drop at each sieve plate of 0.14 atm, or some 280 atm/m to sustain a flow rate of 100 cm/h.

It seems inconceivable that anybody, no matter how vigorous a supporter of the pressure flow concept, could maintain that such an enormous gradient was physiologically possible. Of course, it could be argued that Weatherley and Johnson were applying incorrect values to the size of the filaments and the distance separating them. Tammes, van Die and Ie (1971) have considered the situation in *Yucca*. If the pores are unobstructed, then a velocity of 128 cm/h could be maintained through 1 m of sieve tube by a pressure gradient of 1 atm, a figure not too different from those previously quoted. Using a value of 1000 Å for the distance between filaments in obstructed pores (a value taken from the observation that each pore in *Yucca* is traversed by 10–20 filaments of diameter 250 Å), then the application of Spanner's formula gives a value of only 5.3 atm/m of sieve tube to allow a flow velocity of 44 cm/h (the latter calculated from rates of exudation). While this is much lower than the value derived by Weatherley and Johnson for filament-filled pores, it is still very considerable, and would make it difficult to envisage transport taking place by a pressure flow in plants more than a metre or so in height.

Many people doubt the value of these theoretical approaches to the problem, and there is no doubt that it is possible to vary the supposed parameters of the system *ad nauseam* to arrive at a value which either is or is not compatible with pressure flow. However, provided one is certain that the correct formulae are being applied, it is possible to set the limits at either end of the ultrastructural spectrum, from open to highly occluded pores. Once this has been done, then progress on these lines must inevitably await further discoveries by the electron microscopists.

'Active' mechanisms

Electro-osmosis

The idea that an electrokinetic mechanism could be responsible for long-distance transport in the sieve tubes was put forward independently by two workers—Fensom in 1957 and Spanner in 1958. Fensom's proposals concerning the generation of the electrical gradient at the sieve plates were not very precise; indeed, it now appears that he has discarded electro-osmosis as being of any great importance in solute movement (Tyree and Fensom, 1970). On the other hand, Spanner (1958) has produced concrete proposals as to how the mechanism would function, and he is still firmly committed to electro-osmosis (1970).

In his 1958 paper Spanner reasons that the much-favoured pressure flow hypothesis comes up against severe difficulties from electron microscope evidence on the occluded state of the sieve pores, and that theories based on cytoplasmic streaming fail to satisfy the quantitative requirements of the transport system. Spanner then proceeds to conclude that all the available evidence points to transport occurring as a bulk flow of solution which traverses the sieve plate, all the major components of this solution moving at similar rates. Moreover, Spanner argues that the evidence comes down in favour of movement being an active physiological process, rather than a thermodynamically passive one as suggested by the pressure flow hypothesis.

Having dismissed the two major hypotheses, Spanner then proceeds to examine the possibility that electro-osmosis might fit in with the structure of phloem. The mechanism has three basic requirements: (1) there must be a membrane possessing charged pores; (2) the pores in the membrane must be large enough to allow the free passage of hydrated ions, without being so large that the ions pass out of range of the influence of the fixed charges; and (3) there must be a potential across the membrane, which has to be continuously maintained by metabolic processes in order to allow transport to proceed.

Spanner proposes that the sieve plate itself could form such a membrane. Certainly, if the sieve pores are filled by filaments, which would amost certainly carry fixed charges, then this would provide an excellent system for the generation of electro-osmotic forces. Indeed, filament-filled pores are a prerequisite for electro-osmosis, quite the opposite situation to that necessary for pressure flow. Since the motive force for transport with an electro-osmotic mechanism is generated at the sieve plates, this hypothesis does not encounter difficulties associated with increase in length of the translocation path as is the case with pressure flow, where a twofold increase in path length would reduce the flow rate by half. This consideration, of course, applies to all the active hypotheses, rendering them just as effective in a tall tree as in a small herbaceous plant.

After proposing that the sieve plate fulfils two of the basic requirements for electro-osmosis, Spanner then has to consider how the sieve plate could become polarised. Companion cells were invoked, partly because they lie in intimate contact with the sieve elements, and also because they appear to be the seat of high metabolic potential. It was envisaged that these cells maintain a circulation of a cation (potassium was suggested since it is present in sieve elements in high concentration), across the sieve plate by secretion at one side, followed by uptake at the other with a return path through the companion cells. In a later publication (1970) Spanner has suggested that certain of the 'weaker' sieve elements are forced to conduct in the opposite direction to others, thereby forming a return pathway. The most recent suggestion is for the return path to lie in the apoplast (Spanner and Jones, 1970, *Figure 7.4*).

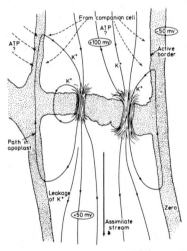

Figure 7.4 Scheme for electro-osmosis through the sieve plate. The potassium ion is actively absorbed above the plate in the presence of high ATP levels. Below the plate it leaks out in the absence of ATP. The electrical potentials are purely hypothetical. (From Spanner and Jones, 1970, courtesy of Springer-Verlag)

Solute movement by electro-osmosis would, of course, be accomplished by a bulk flow of solution. In this it would be very similar to pressure flow, except that there would be no continuous gradient of turgor pressure down a conducting sieve tube. The predicted profile of turgor would have a saw tooth form with a gradient down the lumen of each sieve element followed by a sharp rise at the entrance to the sieve plate. Potassium and sugar concentrations would also have profiles of similar shape (Spanner, 1958).

Apart from the form of these gradients and the continuous expenditure of energy at each sieve plate, the physiological characteristics of solute movement by electro-osmosis would almost certainly be very similar to those of a pressure flow mechanism: simultaneous bidirectional movement could not occur; water, of course, would be transported; and all solutes, with the possible exception of negatively charged molecules, should move with similar velocities. Initiation of transport in a quiescent sieve tube could, according to Spanner (1958), be brought about by the secretion of sugar into the sieve elements at one end—a suggestion which again echoes the pressure flow hypothesis.

At the present time there is little direct evidence for electro-osmosis, although any work which demonstrates movement to be an energy-dependent bulk flow must, of course, put the hypothesis in a favourable light. Apart from the fact that potassium is present within sieve tubes, possibly the most cogent evidence in support of electro-osmosis is the work of Bowling (1968a, 1969). In the first of these publications Bowling attempted to measure the electrical potential difference across the sieve plates of isolated pieces of *Vitis* phloem, immersed in 0.25M sucrose, by inserting microelectrodes at each side of the plate. He found the sieve plates to be polarised with a potential drop of between 4 and 35 mV. The results presented by Bowling (1969) are very favourable to electro-osmosis. He not only confirmed his earlier observations on the polarisation of the sieve plate but also showed that the sieve plates of contiguous elements were polarised in the same direction. In fact, he observed a saw tooth profile of potential, the measurements in one profile being as follows: −43mV, sieve plate, −9mV lumen −36mV, sieve plate, −14mV lumen −29mV, sieve plate, −12mV, i.e. a profile in all essentials of the same form as that predicted by Spanner (1958). The reasonable conclusion reached by Bowling was that his results constituted cogent evidence for electro-osmosis.

There is little doubt that most other workers would have reached the same conclusion, but a note of caution should be added about the interpretation of these measurements, for they were performed, not on undisturbed translocating phloems, but on isolated pieces of tissue. It is always possible that surging of the sieve element contents upon cutting could have led to the pile-up of cytoplasmic material at one side of the sieve plate and its removal at the other. If this material had a fixed charge, which seems most probable, then it would be expected that there would be a Donnan potential at each end of the cell, leading to a potential profile as found by Bowling.

Despite these objections, Bowling's work constitutes the best experimental evidence for electro-osmosis. There have been attempts

(Fensom and Spanner, 1969; Tyree and Fensom, 1970) to determine whether electro-osmosis is a viable proposition, by applying an electrical potential to vascular strands of *Nymphoides* or *Heracleum*, followed by measurements of water and current fluxes. Tyree and Fensom (1970) conclude that their results do not favour an electro-osmotic mechanism in which current flows through a long length of phloem. However, as Spanner (1970) points out, such measurements bear little relevance to his concept of electro-osmosis, in which there is a microcirculation path at each sieve plate.

Despite the manifold attractions of the electro-osmotic hypothesis, there are a number of theoretical objections to the proposal (MacRobbie, 1971) which have not yet been effectively answered by Spanner. The variety of substances transported in the sieve tubes would not appear to be in accord with an electrokinetic mechanism. For maximum efficiency, electro-osmosis should be mediated via ions of one sign only; and, since it is generally considered that the channels in the sieve pores would be negatively charged, the current must therefore be carried by ions of positive sign. If this were the situation, then the system would transport cations and other positively charged molecules, water and neutral molecules, but would preclude any extensive movement of negatively charged molecules. As MacRobbie (1971) has pointed out, many anions such as inorganic phosphate are known to be phloem-mobile, while other solutes (amino acids, organic acids) will be present wholly or partially in a negatively charged form at the pH of sieve tube sap. This objection cannot, of course, be removed by assuming the pores to be positively charged and the current to be carried by anions; this would merely lead to the immobility of the patently high mobile cations such as potassium.

Other theoretical objections to electro-osmosis have also been raised by MacRobbie (1971). Arguing on the basis of the ratio of sucrose to potassium in sieve tube exudates, this author comes to the conclusion that it is most unlikely that one potassium ion could drag more than ten molecules of sucrose through the sieve plate. Thus, from measurements of sucrose mass transfer rates, it is possible to calculate the concomitant potassium fluxes which would be required across the sieve plate and lateral walls of the sieve elements. These have been calculated by MacRobbie to be as high as 40×10^3 pmol cm^{-2} s^{-1}, a value several orders of magnitude greater than those of potassium transmembrane fluxes found in other cells. Even the suggestion of Spanner and Jones (1970) that the plasmalemma of sieve elements is organised into a 'brush border', thereby increasing the effective area for potassium transport, would not, according to MacRobbie, bring down the size of the potassium fluxes required to the values found in other cells.

Spanner (1970) has suggested that a gradient of 25 mV in the electrochemical potential of potassium is required across the sieve plate (a value, incidentally, very similar to that found experimentally by Bowling in 1969). However, MacRobbie (1971) has produced some very cogent arguments which appear to show that the maintenance of a gradient of this size would be beyond the measured energy turnover rates of the phloem. Indeed, Spanner (1970) admits that the greatest difficulty encountered by his hypothesis is the problem of the energetics of sustaining a very high density for the potassium current, although he does not concede that in the present state of our knowledge it is possible to discount the hypothesis on this score.

Current opinion on electro-osmosis may be summed up in the following way. It is a very attractive proposal which in common with the pressure flow hypothesis envisages movement taking place as a bulk flow, yet is in greater accord with what is at present the majority view as to the ultrastructure of sieve elements. It does, however, suffer from a number of theoretical objections, many of which are based on observations with other types of cells. What is urgently required if electro-osmosis is to remain viable are more direct observations on phloem comparable to those of Bowling (1968a, 1969). It should be stressed here that any unequivocal demonstration of simultaneous bidirectional movement must inevitably be disastrous for both the pressure flow and electro-osmotic hypotheses, although in this connection Spanner (1970) appears to think that such would not be the case with electro-osmosis. Possibly Spanner was not thinking of simultaneous bidirectional movement as it has been defined here, but of two adjacent sieve tubes conducting in opposite directions, with lateral exchange of tracers from one to the other. This would produce the pattern of ^{3}H and ^{14}C distribution found by Trip and Gorham (1968a), to which we shall return later.

'Protoplasmic streaming'—cyclosis and transcellular streaming

The suggestion that some form of protoplasmic movement might be responsible for sieve tube transport was advanced by De Vries in 1885. On the face of it, the proposal seems eminently reasonable; circulation of the cytoplasm, i.e. cyclosis, can be observed in many plant cells; surely, therefore, if this occurred within sieve elements, then it must aid in transport. A more recent protagonist of a streaming mechanism has been Curtis (1935), who envisaged movement in the sieve tube lumen being mediated by cyclosis, followed by a diffusional movement across the sieve plates.

There are two major difficulties associated with the cyclosis-type hypothesis: movement across the sieve plates, and the lack of evidence for cytoplasmic movement in mature sieve elements. With regard to the first, it must be clear from what has already been said that thermal diffusion would be far too slow to account for the observed rates of movement. It is also very difficult to conceive that this problem could be surmounted by assuming the sieve plates to be capable of secreting solutes from one side to the other, much in the way that sugars are loaded laterally into the sieve elements, for the specific mass transfer rates along the sieve tube are so great that the energy consumption at each sieve plate would be enormous. It seems that we must reject simple cyclosis as a possible mechanism, although it is worth noting two of the physiological characteristics which would be associated with it: simultaneous bidirectional movement could be accommodated, and there would probably be only small longitudinal fluxes of water in comparison with the solute fluxes.

There is, of course, one way around the problem of how solutes could be moved across the sieve plate in a streaming process, and that is to assume that streaming occurs not only through the sieve element lumen, but also through the sieve pores. This possibility was mooted by Mason, Maskell and Phillis (1936), who were very concerned as to how a process which appeared to show many of the characteristics of thermal diffusion could be 'activated' to transport materials at very high rates. In fact, Mason *et al.* rejected cytoplasmic streaming as a possible 'activating' process on energetic grounds. A number of other workers have also dismissed the idea for the same reason, coupled with the observation that the velocity of cytoplasmic streaming in higher plant cells (a maximum value of 6 cm/h) is very much lower than the commonly observed velocities of phloem transport. This latter consideration does not, of course, necessarily preclude the possibility that motile systems exist in sieve elements, for it can always be argued that these cells might possess mechanisms to produce very high velocities of cytoplasmic movement. In this connection, MacRobbie (1971) has quoted work by Kamiya (1953) on streaming in the slime mould *Physarum*, where velocities up to nearly 500 cm/h have been demonstrated.

There is no doubt that the interest of transport physiologists in some form of 'cytoplasmic' mechanism has been growing in recent years, discussion of the subject having been largely revived by the work of Thaine (1961, 1962, 1964). Reference has been made in the previous chapter to transcellular strands; certainly, if these exist and are responsible for movement, then they would obviate the difficulties encountered by cyclosis at the sieve plate, for transcellular strands would pass through the sieve pores.

In Thaine's exposition of his hypothesis (1964) he proposed that 'transcellular strands form parallel pathways for mobile materials moving across sieve tube lumina, and that these strands penetrate the plasmalemma and the sieve plates to pass through sieve pores'. The strands here would be static, transport through them being possibly brought about by the movement of microscopically visible moving particles. The functioning of these strands, according to Thaine (1964), does not require a quantitative relationship between the rate of movement and the steepness of a solute concentration gradient, although the concept of solute potential could be applied to the loading and unloading of strands which would lead to movement from regions of high solute concentration to regions of low solute concentration, i.e. from sources to sinks.

Although Thaine was the originator of the transcellular streaming hypothesis, other workers, notably Canny (1962) and Canny and Phillips (1963), have modified Thaine's original ideas and have attempted to put these modifications on a quantitative basis. In the first of these publications Canny's premises are set out as follows. Transcellular strands are assumed to be of general occurrence in sieve tubes and to occupy up to half the cross-sectional area of a sieve element. The space between the strands acts as a reservoir for a solution of sucrose and other solutes, and it is this reservoir which is tapped by aphid stylets and incisions. The strands require to be membrane-bound, and, unlike Thaine's model, the whole of each strand moves in a type of bulk flow.

Canny (1962) has given a diagram, reproduced here as *Figure 7.5*, which illustrates the working of his model. In a sieve tube which is not transporting, the strands would be streaming in both directions at equal velocities, the sugar concentrations in all the sieve element reservoirs also being equal. If sugar is now removed from an element at the sink end of the system, the concentration in that reservoir will fall, which will lead to transfer of sugar from the next element. Canny visualises a 'wave of adjustment' as passing along the sieve tube at a velocity slightly less than the rate of streaming, leading to the formation of a concentration gradient from source to sink when a steady state is attained; the greater the rate of sugar removal at the sink, the steeper will be the gradient. The rate of transport is then proportional to the steepness of the gradient, in much the same way as in a simple diffusion system. According to Canny (1962), the only difficulty which is encountered by his model in explaining the known characteristics of phloem transport is that concerned with the problem of the very high rates of mass transfer known to exist.

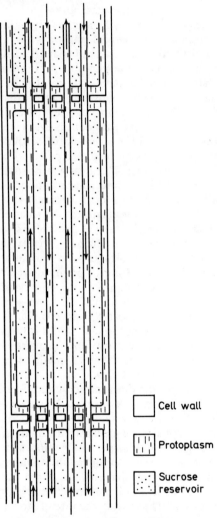

Figure 7.5 Transcellular streaming mechanism. Diagram of the proposed structure and movement within a sieve tube. (From Canny, 1962, courtesy of The Clarendon Press)

In both the paper by Canny (1962) and also the detailed analysis of the model given by Canny and Phillips (1963), transport is envisaged as being brought about by the movement of a concentrated solution, impelled through the strands by cytoplasmic streaming. However, a streaming mechanism runs into certain difficulties, as pointed out by Weatherley and Johnson (1968). For instance, how do these streaming strands 'turn round'? This process must presumably occur at some point, otherwise there would be a 'pile up' of cytoplasmic material at each end of the system. This difficulty would not, of course, be encountered in the system of Thaine (1964), where the strands themselves are stationary, although any moving particles within the strand which were not concerned with solute transport would presumably have to turn round. Another difficulty with Canny's system is that if the strands in one sieve tube stream in opposite directions simultaneously, there would have to be a large concentration difference between the solutions in the up and down strands in order to produce a net flux of sugar in one direction. It is very difficult to visualise how this could occur, for surely it would lead to large water fluxes between the strands and the reservoir. It might be necessary to assume different solute permeabilities for the up and down strands, a condition which appears unlikely.

The problem of strand 'turn round' can be removed if a peristaltic rather than a streaming mechanism is proposed. Indeed, it was Thaine (1969), followed by Jarvis and Thaine (1971), who first proposed the idea that peristaltic contractions might occur which could drive a solution through transcellular strands. Thaine (1969) visualised protein filaments within the transcellular strands which, by contraction and relaxation, could produce a peristaltic wave along the strand. This would be capable of pumping the fluid contents of the endoplasmic tubules within the strands over long distances and at high velocities.

The concept of peristaltic action has been quantitatively analysed by Aikman and Anderson (1971), employing a simplified transcellular strand system. These workers have come to the conclusion that peristalsis is quantitatively acceptable, although to achieve the highest mass transfer rates (i.e into fruits) the strands would have to occupy an appreciable proportion of the phloem cross-section, and would virtually all have to pump in the same direction. In organs such as petioles which exhibit relatively low specific mass transfer values, it would be possible for some strands to pump in one direction while others pumped in the opposite sense.

On the analysis of Aikman and Anderson, a net flux in one direction could be produced either by concentration differences between the up and down strands or by differences in their relative numbers. They believe that they have obviated the problem of large water fluxes between strands running in opposite directions, by postulating a turgor pressure balancing of the osmotic driving forces.

Of all the mechanisms considered so far, with the exception of cyclosis and surface movement, transport in discrete transcellular strands is the only one which would allow 'true' simultaneous bidirectional movement. However, strand movement certainly, if brought about by peristalsis, could differ from cyclosis in the context of water movement. While it seems probable that cyclosis would not involve large longitudinal water fluxes, there is no question at all about peristaltic movement, for this would be mediated by a flow of solution.

There is little doubt that the transcellular strand hypotheses embody a number of attractive features: i.e. a plausible explanation of simultaneous bidirectional movement (if this is unequivocally shown to exist), and a reason for certain observations showing that the phloem can have a high metabolic rate (although this consideration would also apply to any of the other active hypotheses). However, it can also be said that these hypotheses rely on a degree of speculation which is quite unacceptable to many workers.

Canny (1962) bases much of his argument on the so-called logarithmic profile of tracer distribution down a stem from a source, believing this to be produced within the sieve tubes, the form of the profile being a reflection of the nature of the mechanism. A much more plausible explanation could be found in the analysis of Horwitz (1958), who showed that a tracer distribution of this form could arise by 'leakage' from the sieve elements into other cells. Canny does not believe that leakage takes place to any marked extent, a view difficult to reconcile with the function of the phloem as a distribution system (Crafts and Crisp, 1971).

Clearly, the detailed mathematical analyses of Canny and Phillips (1963) and of Aikman and Anderson (1971) could be quite futile if the cytological keystone of their work were proved to be lacking. As we have already seen, the existence of membrane-bound strands has not been proved, the great majority of cytologists being firmly opposed to the proposal.

Other 'cytoplasmic' hypotheses, contractile filaments

While there are few who would support the existence of transcellular strands, many believe in filamentous constituents of sieve elements. What, then, is the function of such filaments? To those who consider them to be an essential feature of functioning elements the question is of prime importance. Spanner, as we have seen, has implicated them in his electro-osmotic mechanism, but there are many who do not share his views. Quite recently, it has been suggested by Weatherley and Johnson (1968), by MacRobbie (1971) and by Fensom (1972) that the filaments might constitute a contractile system, by means of which a bulk flow of solution could be propelled along the sieve tubes.

MacRobbie (1971) bases her arguments on the observations of a number of workers on cytoplasmic streaming in the alga *Nitella* and in the slime mould *Physarum*. In *Nitella* streaming of an inner layer of cytoplasm occurs, the motive force for which is generated at the boundary of this inner layer with an outer gel layer (Kamiya, 1959; Kamitsubo, 1966). It is possible to extrude drops of cytoplasm from the cells, and observation of these (Jarosch, 1964) has revealed the presence of fibrils which undergo movement. It has been suggested by various authors (Kamiya, 1960) that these fibrils are able to generate the forces necessary to produce streaming. With the electron microscope (Nagai and Rebhun, 1966) bundles of filaments can be seen, each made up of microfilaments some 50 Å in diameter. In *Physarum* there is also a considerable weight of evidence that fibrils are associated with streaming (Kamiya, 1968), and these fibrils may have close relationships with the muscle proteins actin and myosin.

From these observations it seems a simple and not unreasonable extrapolation to suggest that the P-protein filaments found in sieve elements are akin to those associated with streaming in other cells and may therefore form a contractile system. Certainly, the P-protein filaments have a similar diameter to that of the fibrils in *Nitella* and the slime moulds. Recent work (Kollmann *et al.*, 1970) has also indicated the P-protein filaments to have a beaded appearance, somewhat similar to the 'actin' filaments of other cell types, e.g. in *Acanthamoeba* (Pollard *et al.*, 1970). Kleinig *et al.* (1971) have studied certain of the biochemical characteristics of the proteins in exudates from *Cucurbita*, in some of which (reversible aggregation, precipitation by calcium ions) they are similar to actin.

It seems, according to MacRobbie (1971), that if a system of polarised actin filaments exists in sieve elements (polarisation could be achieved by the establishment of a turgor gradient along the sieve tube), then these would probably be able to generate sufficient force to drive a solution through the elements. This conclusion was reached by taking the figure of Kollmann *et al.* (1970) for the concentration of P-protein in *Cucurbita* exudate (10 mg/ml). Assuming this to consist of 80 Å diameter filaments, their total lengths would be 2×10^{10} cm. Knowing the force-generating capacity of *Nitella* filaments, MacRobbie calculates that if the P-protein filaments are comparable, they should be capable of producing a force of approximately 2 atm/m, i.e. a pressure gradient of sufficient magnitude to move a solution at the velocities which are evident in phloem.

A rather detailed exposition of a 'contractile filament' hypothesis has been produced by Fensom (1972) subsequent to his rejection of electro-osmosis as a major mechanism responsible for transport. Part of the evidence adduced by Fensom to support his thesis comes from work by Lee, Arnold and Fensom (1971), using Nomarski interference optics to observe living sieve tubes in *Heracleum* phloem strands. 'Marker particles' were observed which executed saltatory movements (Brownian motion was dismissed as being responsible), it being presumed that these particles were attached to filaments and that the particle movement therefore reflected movements of the filaments. Small particles were also seen 'bouncing' close to sieve plate pores, but they did not apparently pass through the latter, an observation which Lee and his colleagues interpreted as showing that the pores were blocked by fibrillar material.

Fensom (1972) elaborates his arguments, using data obtained from experiments in which tracers ([^{14}C]sucrose, ^{42}K and ^{3}HHO) were fed to phloem strands of *Heracleum* either by microinjection (Fensom and Davidson, 1970) or by surface applications. With the first method, ^{14}C was found to move in both directions from the injection site, activity being present along the strands in the form of discrete pulses of activity. When ^{42}K was injected simultaneously with ^{14}C, it usually remained within 3 cm of the injection point. It was inferred from these results that the two tracers are not transported by the same mechanism. Surface application of tracers produced essentially similar results, the ^{14}C-activity again appearing in pulses at a distance from the application point. ^{3}H from ^{3}HHO also moved in the form of pulses, although not closely associated with those of ^{14}C.

Fensom (1972) postulates that the fibrillar system could exist as a flexible network of helices with hollow centres. The walls of the fibrils would be capable of generating contractile waves moving in an axial direction (*Figure 7.6*). He proposes that the waves could carry 'packets' of sucrose and amino acids at high concentrations in the fibrils, the pulse distribution of [^{14}C]sucrose referred to previously being produced by the loading of these fibrils in a pulse-like manner. A further consequence of the proposed system would be that a mass flow outside the fibrils could be produced by the contractile waves. In this flow [^{14}C]sugars, ^{42}K and ^{3}HHO would move together.

Essentially, then, Fensom postulates that at least two mechanisms are involved (bimodal), and possibly three, since he does not discount some surface movement. Since one of the modes is a bulk flow, water as well as solutes must move, but this does not rule out the possibility of simultaneous bidirectional movement associated with one of the other two postulated mechanisms. Fensom believes that he can justify his proposals on quantitative grounds, particularly with regard to the essential features of known specific mass transfer rates and energy dissipation.

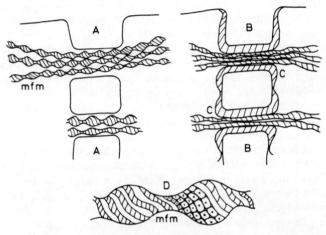

Figure 7.6 Diagram of possible arrangement of fibrillar material passing through sieve plates. A–A, microperistaltic waves in phase allowing mass flow outside fibrils, no callose. B–B, callose deposits. C–C, constructing sieve pores in a cut phloem strand, preventing mass flow and restricting pulse flow. D, enlargement of a possible structure for the fibrils to allow a contractile protein mechanism to operate, left, by adjacent sliding of wall or, right, by beaded units which enlarge by flattening and contract by thickening. mfm, microfibrillar material (phloem protein fibrils). (From Fensom, 1972, courtesy of the National Research Council of Canada)

It is the proposition that several mechanisms are involved in solute transport which constitutes one of the most attractive features of Fensom's hypothesis, since it enables us readily to reconcile many of the contradictory results from physiological investigations on phloem which will be reviewed in subsequent chapters. While Fensom's specific proposals are new, the idea that solute transport could be achieved by more than one mechanism is not: Arisz suggested the possibility in 1952.

In this and the preceding chapter we have examined current opinion on the fine structure of the sieve elements and the hypotheses put forward to explain the mechanism of movement. In viewing the whole picture, one can hardly feel optimistic about a substantial degree of progress in the unification of ideas on mechanisms in the immediate future. Pressure flow was, and remains to some physiologists, an extremely attractive proposition. However, the reasons which lay behind the formulation of the hypothesis were almost totally physiological (exudation phenomena), not structural. In the early years of ultrastructural investigations it was relatively easy to find support for the hypothesis, but the recent trend has been for investigators to show the sieve element as a cell with a highly complex structure, apparently quite incompatible with a flow driven by osmotic pumps at each end of the system.

Such is the confusion in the understanding of phloem that it is not possible to rule out pressure flow merely on the basis of the application of certain laws of physics to the structure as observed with the electron microscope. It is not beyond the bounds of possibility, as Weatherley and Johnson (1968) have indicated, that there may be a basic error in our reasoning. It could well be argued that the physical principles which we apply to movement in sieve tubes are inadequate to meet the situation in such small conduits. Is it not possible, for instance, for the frictional resistances to flow in sieve tubes to be very much lower than are stated by the Poiseuille equation, thereby enabling a solution to move without the provision of physiologically enormous amounts of energy?

The great problem with this latter type of reasoning lies in the temptation to carry on the process *ad infinitum*, thus leading to an enormous inverted pyramid of assumptions which bears little relevance to established facts.

8

Metabolic energy and transport

As we saw in the previous chapter, the proposed mechanisms of movement can readily be assigned, depending upon their relationships to metabolic energy, to either an active or a passive group. It is, therefore, not surprising that phloem physiologists have for some considerable time been striving towards an elucidation of the energy requirements of the transport system. Broadly speaking, there have been two types of approach to the problem, neither of which has yet led to any unanimity of opinion.

One method of attack has been to try and measure what could be termed the 'metabolic potential' of the phloem, i.e. to try and answer the question as to whether the tissue is or is not capable of high rates of metabolic activity. The types of experiment which have been executed in pursuit of this aim include those designed to measure the respiration rate of phloem, the enzymic content of phloem saps and the metabolism of ATP.

It seems frequently to be assumed that if phloem is shown to possess a high metabolic potential, then this would be indicative of the operation of an active transport mechanism. However, if a little thought is given to the matter, it becomes clear that such an inference does not necessarily follow. There is abundant evidence for the operation of solute-loading and -unloading mechanisms not only at sources and sinks but also along the whole transport path (see the papers of Weatherley *et al.*, 1959, and Peel and Weatherley, 1962). Could it not be, therefore, that a high metabolic rate might be associated with these processes rather than with longitudinal movement? All that can be justifiably attempted at the moment is to measure respiration rates of phloem and then to use these rates in calculations with the object of determining whether a particular mechanism is energetically feasible.

Just as it is not proper to argue directly from a metabolically active phloem to an active longitudinal transport mechanism, it is also unjustified to draw firm inferences in the opposite sense. Those observations which seem to show that the sieve elements are metabolically moribund (for instance, the reports of mitochondrial degeneration in mature sieve elements referred to in Chapter 6) do not in themselves necessarily favour the adoption of a passive mechanism. As we have noted, intimate connections exist between the sieve elements and companion cells; therefore it seems very possible that the latter cells could provide the necessary energy to power an active mechanism, particularly as there appears to be no dissension from the view of the companion cells as a metabolic 'powerhouse'.

Another type of approach to the issue of the connection between metabolic energy and longitudinal transport is the use of conditions (low temperature, inhibitors and anoxia) which reduce or cause complete cessation of metabolic activity. Clearly, it is no use in this type of experiment to subject the whole of the translocation system to a metabolic depressor, since this would certainly interfere with the energy-consuming processes involved in solute loading. It is therefore necessary to confine the treatment to a small portion of the path length, any effect on transport being monitored by, for instance, following radiotracer movement along the path. On the face of it, it seems that this type of experiment should provide a clear resolution of the problem; if low temperature, etc., cause a decline or stoppage of transport, then surely transport must be active. However, the results of this work are, as we shall see, not easily interpreted, since many factors have to be considered which could affect the data extracted from these experiments.

The 'metabolic potential' of phloem

Rates of respiration

Experiments on the respiration rates of phloem have almost always been performed on isolated pieces of tissue, being generally designed to answer two questions: (1) What is the magnitude of the process? (2) Does the rate of respiration differ markedly from that of ground parenchyma tissue? Typical of these investigations is the work of Duloy and Mercer (1961) on *Cucurbita* and certain other species. With *Cucurbita*, tissue was obtained by taking the first three mature internodes of a runner, cutting these into 1 cm lengths and then splitting the pieces vertically into two. The bundles of the inner ring were dissected out while some of the remaining tissue was trimmed so as to include only parenchyma. Respiration rates were measured with the tissues immersed in phosphate buffer.

Measurements of the respiration rates of phloem tissues, taken 3 h after these had been excised from the stem, showed that the oxygen uptake of the vascular bundles was some three times greater than that of the ground parenchyma, when compared on a weight for weight basis.

Moreover, if the percentage of tissue other than phloem in the vascular bundles was taken into account, it was calculated that phloem respired at a rate up to 15 times that of the ground parenchyma. In another species studied by Duloy and Mercer, *Apium graveolens*, calculated values of phloem respiration rate were found to be as high as 2000 μl O_2 per gram fresh wt. per hour, while that of the parenchyma only reached 30 μl O_2 per gram fresh wt. per hour.

Duloy and Mercer extended their investigation in two directions, viz. to find whether any qualitative difference existed between respiration in phloem as contrasted with parenchyma, and to discover the reason for the high rates found in phloem. They were unable to detect any differences in the pathways concerned in the two tissues, arriving at the conclusion that both were probably mediated via the tricarboxylic acid cycle. Measurements of the protein nitrogen content of the two tissues revealed the phloem to contain up to ten times the quantity of protein per unit fresh weight when compared with parenchyma tissue. Thus, they inferred that the higher respiratory activity of the phloem was the result of its having more cytoplasm per unit volume than the parenchyma.

There are a number of other references in the literature showing vascular tissues to have a higher respiration rate than parenchyma tissues. *Table 8.1* gives a list of those quoted by Kursanov (1963). However, there are other studies, such as that by Canny and Markus (1960), working on *Vitis vinifera*, in which the conclusion was reached that phloem does not have a respiration rate significantly higher than that of parenchyma tissue.

It is always possible that the high rates found in excised phloem might well be produced as a result of injury to this sensitive tissue. (Duloy and Mercer, 1961, found that the rate of respiration dropped markedly with time.) Canny (1960a) has, however, developed a method for measuring the rate of breakdown of ^{14}C-labelled sugars during their transit through intact organs. Petioles of *Vitis* were enclosed in a gas-tight respiration chamber, the lamina attached to the petiole being allowed to assimilate [14]CO_2. By assaying the [14]CO_2 produced by the part of the petiole enclosed within the chamber, Canny was able to measure the rate of $^{14}CO_2$ output from labelled sucrose within the phloem (he produced autoradiographs showing that the ^{14}C-label was confined to this tissue).

Table 8.1 RESPIRATION RATES OF LEAVES, VASCULAR BUNDLES AND PARENCHYMA OF THE PETIOLES OR STEMS OF VARIOUS PLANTS. (FROM KURSANOV, 1963, courtesy of *Academic Press*)

Plant	Respiration rate, $\mu l\ O_2\ g^{-1}$ *fresh wt.* h^{-1}			
	Bundles	*Leaves*	*Petioles* (*minus bundles*)	*Reference*
Sugar beet	572	416	100	Kursanov and Turkina, 1952
Sugar beet	462	232	95	Tsao Tsng Hsun and Lui Chih-Yi, 1957
Plantago major	820	374	228	Kursanov and Turkina, 1952
Plantago major	500	—	146	Willenbrink, 1957
Primula leesiana	309	—	32	Willenbrink, 1957
Heracleum mantegazzianum	230	—	32	Ziegler, 1958
Cucurbita pepo	540 (phloem only calculated)	—	60	Duloy and Mercer, 1961

The values obtained for CO_2 evolution from the phloem were found to be around 200 μl per gram fresh wt. per hour, a figure somewhat lower than those given by other workers for oxygen consumption of isolated tissues. As Canny himself says, these measurements hardly support the concept of the phloem as a tissue with a high metabolic activity; indeed the figure quoted was considerably less than the CO_2 evolution rate from the whole of the petiole. However, Canny was, of course, measuring only that part of the phloem respiration which was utilising the labelled sucrose as a substrate. It could well be that a considerable part of the whole petiole respiration might have been contributed by the breakdown of unlabelled carbohydrates in the phloem, present in this tissue before the start of experimentation. It must therefore be concluded that the figures given for phloem respiration by Canny are minimum values only.

Having thus arrived at some sort of estimate of the respiratory potential of phloem, we must now try and determine whether the measured rates of sugar breakdown are sufficient to support the various proposed mechanisms. In attempting to evaluate the situation, we immediately place ourselves in a quandary, for we are completely in the dark about what proportion of the total respiration might be used to power the translocation mechanism.

Canny (1960a), as well as providing measurement of the rate of respiratory breakdown in intact vine petioles, has also given a degree of consideration to the problem of the energy requirements of the transport system. In certain of his experiments, Canny measured the ratio (Q) of amount of ^{14}C-labelled carbon dioxide given off from a petiole during a given period of time to total radio-activity contained within the sugars subsequently extracted from the organ. The magnitude of this ratio effectively provided a measure of the instantaneous sucrose content which was lost by respiration during the experimental period. In fact, Canny found a remarkable degree of constancy in the values of this ratio measured on a number of petioles, depite the fact that the sucrose extracted from the tissue varied widely in its specific activity. This, according to Canny '. . . is suggestive of an orderly rather than a random process . . .', and he used the constancy of the data as an argument in favour of taking the whole of the energy provided by the respiratory breakdown of the tranlocated sugar as being available to drive the longitudinal transport system.

However, these assumptions have been questioned by Coulson and Peel (1968), who obtained a series of values for this ratio from 3–5-week-old shoots and 2–3-year-old stems of willow using a comparable technique to that of Canny (1960a). Certainly, Coulson and Peel found a reasonable constancy of the ratio Q, the values being higher in the young shoots than in the old stems, although there appeared to be a greater degree oi variability than in Canny's data. What Coulson and Peel did not find was any support for the concept that a considerable proportion of the energy produced by the respiration of the transported sugars is utilised to power phloem transport.

It seems most probable, therefore, that much of the energy derived from sugar breakdown may be used for purposes other than transport. Certainly, if this were so, it would explain the higher Q-values in young, rapidly growing willow shoots, when compared with those obtained from older, less active stems. In view of the lack of data on the total quantity of energy available for translocation, all that can be done is to take what appears to be a reasonable figure for the rate of phloem respiration and then to presume that all the energy produced as a result of this is channelled into transport. Admittedly, this is a highly unsatisfactory compromise, although probably not more so than the application of formulae to flow in sieve tubes which we have reviewed in the previous chapter. At all events, by taking a maximum value for the available energy, it should be possible to decide which mechanisms are energetically feasible and which are not.

Before any calculations can be made, we have to decide what figure should be taken for the respiration rate of phloem. There is little justification for assuming a value as high as the unsubstantiated ones reported by some of the Russian workers. On the other hand, the figure of 200 µl CO_2 per gram fresh wt. per hour, representing a glucose breakdown rate of 0.012 g glucose per cubic centimetre of phloem per day (Canny, 1960a) is somewhat lower than the values given by a number of workers (*Table 8.1*). A figure of 0.05 g glucose per cubic centimetre of phloem per day, as used by MacRobbie (1971) and Weatherley and Johnson (1968), would not appear to be unreasonable, representing an oxygen uptake rate of around 500 µl per gram fresh wt. per hour.

Pressure flow

Assuming respiration to be 100% efficient and all the available energy to be used in pumping. Weatherley and Johnson (1968) have made the following calculations. If 1 cm² of sieve tube is considered with a pressure drop of P atm through each sieve element and the velocity of transport is V cm/h, then the work done is

$$P \times V \times 10^6 \text{ ergs per hour per sieve element}$$

Thus, if there are n sieve elements per centimetre, then the work done per cubic centimetre of sieve tube is

$$nP \times V \times 10^6 \text{ erg/h}$$

which, divided by the mechanical equivalent of heat, converts to calories, viz.

$$\frac{nP \times V \times 10^6}{4.2 \times 10^7} \text{ cal/h}$$

Hence, if the free energy released from the combustion of glucose is known, the glucose breakdown rate required to maintain a given pressure gradient can readily be calculated.

It becomes evident that if the sieve pores are open, then a glucose consumption of 0.05 g per cubic centimetre of phloem per day would be more than adequate to propel a pressure flow, for, assuming a value for V of 100 cm/h, this rate of respiration could maintain a pressure gradient of 3.3 atm/cm. However, if the pores are occluded by filaments and resistance is met with in the lumen, then the operation of a pressure flow mechanism seems most unlikely. In the previous chapter calculations using a formula derived by Spanner were quoted in which a pressure gradient of 280 atm/m was necessary for flow at 100 cm/h through sieve plates having pores filled with filaments 100 Å in diameter and lying 200 Å apart. Such a gradient could just be maintained by the respiration rate chosen. However, if the lumen of the sieve element was also filled with filaments having the same size and configuration, then the respiration rate required would have to be two orders of magnitude greater than the standard figure.

When it is considered that respiration operates at an efficiency far below 100%, it becomes evident that pressure flow is only energetically possible if the lumina of the sieve elements are unobstructed and if the sieve pores, at most, only contain widely dispersed filaments.

Electro-osmosis

Although Spanner (1962) was originally of the opinion that sufficient energy was available from respiration to power an electro-osmotic mechanism, it is clear that he has become more cautious recently on this point. As mentioned in the previous chapter, MacRobbie (1971) has produced some calculations on the energy requirements of electro-osmosis which throw some doubt on the value of this hypothesis.

Taking the suggestion of Spanner (1970) that a potential gradient of 25 mV per sieve plate is required, MacRobbie derives a figure of 0.59 kcal/mol to raise the electrochemical potential of potassium by this amount. From the free energy of hydrolysis of ATP, the maximum number of potassium ions which can be pumped per molecule of ATP is 20; thus it is possible to derive a figure for the required ATP consumption in phloem. If petioles are considered (measured mass transfer rates are relatively low in these organs), then there must be a potassium flux of 0.25×10^6 pmol $cm^{-2} s^{-1}$ through each sieve plate. Through 1 cm^2 of phloem, a fifth of which consists of sieve tubes (this may be far too low an estimate by MacRobbie; recent measurements on tree phloem have shown the proportion to be as high as three-quarters—Lawton and Canny, 1970), a potassium flux of 5×10^4 pmol/s would be required. If there are 20–100 sieve plates per centimetre, then in 1 cm^3 of phloem there would have to be a total potassium recirculation of $(1-5) \times 10^6$ pmol/s, necessitating an ATP turnover rate of $(5-25) \times 10^4$ pmol $s^{-1} cm^{-3}$.

A glucose breakdown rate of 0.05 g per cubic centimetre of phloem per day would enable an ATP production rate of 11×10^4 pmol/s to be maintained, a figure twice that necessary if the sieve plates were 500 μm apart, but not nearly sufficient if the sieve elements were only 100 μm in length. MacRobbie makes the valid points that the situation would be much worse in fruit stalks, where the highest mass transfer rates are found, and would be exacerbated further if only a limited amount of the total energy production was available for transport.

Streaming and motile mechanisms

One of the earliest calculations of the energy requirements of a particular mechanism was that given by Mason *et al.* (1936) for cyclosis, the conclusion being reached that this proposal was energetically untenable. In subsequent assessments by Palmquist (1938) and Canny (1961) it was suggested that Mason and his colleagues had made an error in the velocity, leading to a 400-fold error in the energy requirement. Further calculations by Spanner (1962) have shown the estimates of Mason and his colleagues to be correct, and therefore cyclosis can apparently be discarded.

Movement of a solution by pressure flow through transcellular strands of 5 μm diameter occupying half the cross-sectional area of the sieve element would, according to Weatherley and Johnson (1968), be quite possible with a glucose breakdown rate of 0.05 g per cubic centimetre of phloem per day. Aikman and Anderson (1971), in an analysis of a simplified version of the peristaltic model of Thaine (1969), assume efficiencies of 50 % for ATP production from sucrose and 25 % for the conversion of the chemical energy of ATP to mechanical work. Calculations of the required sugar consumption then lead to a figure, acceptable in the context of the standard figure and the above provisos, of 0.04 g per cubic centimetre of phloem per day. It must, however, be emphasised that Aikman and Anderson analysed a system in which strands consisted only of a simple, membrane-bound tube. If the structure of these was as complicated as suggested by Thaine (1969), the energy requirement could be much greater than the figure given above.

To summarise the opinions of various authors on the energetics of transport, it seems that pressure flow in unobstructed sieve tubes is readily possible, but the presence of a network of filaments, unless this is extremely sparse, must put pressure flow beyond the bounds of credibility. Electro-osmosis on Spanner's present formulation is also on the borderline of acceptability, while cyclosis must be dismissed. Peristaltic movement in strands or in microfibrils seems, however, to be just possible.

Sites of respiration in phloem

It frequently seems to have been assumed by supporters of the pressure flow hypothesis that any observations which show the functioning sieve element as a metabolically inactive cell must necessarily favour this hypothesis. While, within certain limits, this cannot be denied, it does not necessarily follow, as has already been indicated, that sieve elements have to be capable of respiratory activity to sustain an active mechanism; the energy could readily be provided from associated cells. Thus, any arguments based on the respiratory capabilities of sieve elements are to a large extent sterile, and this, and the technical difficulties associated with the investigation of the problem, appear to have blunted the zeal of phloem physiologists in this direction.

An approach to the investigation of the respiratory capabilities of sieve elements has, however, been devised by Coulson and Peel (1968), the principles underlying their experiments being as follows. Colonies of aphids were sited on stems of willow, the purpose of these being to directly monitor changes in the specific activity of the sieve tube sap after allowing the leaves to assimilate labelled carbon dioxide. A chamber was sealed to the stem immediately in front of the aphid colony in order to measure the rate at which labelled carbon dioxide was produced by the respiratory breakdown of the labelled sugars in transit through the phloem. It was argued that if the labelled sugars were respired within the sieve elements, then the slopes of the increase in the sieve tube sap and respiratory gas activities with time should be the same. On the other hand, if respiration occurred in cells adjacent to the sieve elements, then it might be expected that the respiratory gas activity should be reduced by 'dilution' of the labelled sugars with unlabelled sugars already present. (It should be noted that this argument may not apply to the companion cells, for it is unlikely that they carry much in the way of sugar pools.)

The results of this study were equivocal. However, some of the data indicated that in 3–5-week-old shoots, at least, some respiration of the labelled sugars was taking place in the sieve elements.

Pathways of respiration in phloem

Duloy and Mercer (1961) have performed experiments to determine the characteristics of the respiratory pathway in the phloem and parenchyma tissues of several species. It was shown that the R.Q. values for both tissues in water were close to unity (R.Q., i.e.

respiratory quotient $= CO_2$ evolved in respiration/O_2 absorbed), and the addition of sucrose to the tissues produced little or no change in the value of the quotient. These observations were taken as good evidence for the assumption that the primary substrate for respiration was a soluble carbohydrate.

Further studies with intermediates of the tricarboxylic acid cycle (*Table 8.2*) showed the respiration rate of phloem and parenchyma to be substantially enhanced by the addition of α-ketogluterate, succinate and malate, while the R.Q. values were little affected. Moreover, the addition of 0.02M malonate produced a 70-95% inhibition in the oxygen uptake rate of both tissues within 1 h from application, unless certain of the tricarboxylic acid cycle intermediates were present.

Table 8.2 PERCENTAGE INCREASE IN OXYGEN UPTAKE BY PHLOEM AND PARENCHYMA TISSUES FROM *Cucurbita pepo* PRODUCED BY INTERMEDIATES OF THE TRICARBOXYLIC ACID CYCLE. (FROM DULOY AND MERCER, 1961, courtesy *C.S.I.R.O.*)

	Malate	*Succinate*	*α-ketoglutarate*
Vascular tissue in buffer	73	62	80
Vascular tissue in buffered sucrose	15	17	27
Parenchyma tissue in buffer	39	92	37
Parenchyma tissue in buffered sucrose	31	20	27

Duloy and Mercer came to the conclusion that the respiratory pathway in phloem and parenchyma was essentially similar to the tricarboxylic acid cycle, and that no apparent qualititative differences existed between the two tissues.

Enzymes in sieve elements

At first sight, it might appear that a study of the enzymic contents of sieve tube saps could provide a good indication of the metabolic capabilities of these cells. It was this supposition which led Wanner (1953) to investigate the range of enzymes present in exudates from *Robinia* phloem. The apparent paucity of respiratory enzymes (phosphoglucomutase was the only one of the glycolytic enzymes detected) led this worker to conclude that the sieve elements were metabolically inert.

Subsequent studies have, however, to some extent reversed this view, at least as far as the number of enzymes present is concerned. Lester and Evert (1965), using a histochemical technique involving the precipitation of lead sulphide, showed acid phosphatase to be present, not only in the companion and parenchyma cells of *Tilia* phloem but also in the sieve elements, where the enzyme was associated with a system of internal strands.

More recently, Kennecke *et al.* (1971) have re-examined the enzymatic spectrum of *Robinia* saps. Despite the use of highly sensitive tests, no invertase was detected, a result which confirmed the earlier findings of Wanner (1953). As Kennecke *et al.* point out, it is not surprising that this enzyme is absent, for it would rapidly degrade the sucrose during transit. Moreover, since the phloem tissues as a whole contain abundant invertase, the absence of this enzyme in the sap indicates that little if any contamination of the exudate occurred from the contents of other cells. Thus, any enzymes detected in the exudate must be present in the intact sieve elements.

Kennecke *et al.* (1971), in fact, demonstrated the exudates to contain a wide variety of enzymes, including most of those concerned with the glycolytic pathway, and the interconversions of starch ⇌ glucose. Certain of the enzymes of the tricarboxylic acid cycle also showed activity (isocitrate dehydrogenase, fumarase and malate dehydrogenase), while some were present which are involved in connecting the products of carbohydrate breakdown and nitrogen metabolism, thus providing the possibility of protein synthesis in sieve elements. Eschrich, Evert and Heyser (1971) were able to detect peroxidases, acid phosphatases and aldolases in *Cucurbita* exudate, although they were unable to find some of those concerned with carbohydrate metabolism.

It would thus seem that we must now look upon the sieve elements as potentially highly active cells in a metabolic sense, particularly as the studies of Kennecke and his colleagues did not include investigations of the enzymic content of the peripheral cytoplasm. (Callose synthetase is almost certainly present here, judging from the speed at which callose can be deposited.) However, although the *potential* of enzymic activity in these phloem exudates is high, there is considerable doubt as to whether many of them act within the milieu of the unaltered exudate. For instance, the temperature, ion and sucrose contents of the test solutions employed by Kennecke *et al.* (1971) to demonstrate the presence of the enzymes were quite different from those of the exudate; furthermore, the pH optimum of many of the enzymes was well below the high values normally found in phloem saps.

ATP in sieve tubes

There are a number of references in the literature showing ATP to be present in sieve tube exudates at high concentrations. Kluge and Ziegler (1964) analysed exudates from a variety of species, recording concentrations within the range 0.03–0.30 μg/μl. More recently, Gardner and Peel (1969) have found ATP in sieve tube exudate, obtained from willow via the aphid stylet technique, at concentrations as high as 1 μg/μl.

However, just as with the reports of the presence of enzymes, we cannot argue from these findings on ATP levels to the conclusion that the sieve tube is a metabolically active cell, for we have as yet no knowledge of the function of this substance; it may be used to power the translocation machinery or it may be required for some other purpose. One thing seems to be certain, however: ATP is a normal constituent of the sap, for it is not merely present as a result of puncturing the sieve element. (ATP secretion rates do not fall markedly with time—Gardner and Peel, 1969.) Also, ATP can be readily transported through the sieve tubes (Gardner and Peel, 1972b).

Some evidence which can be adduced in favour of the contention that the ATP in sieve tubes is employed to drive a translocation mechanism comes from the work described in certain recent publications. A rapid turnover of the ATP is indicated by the speed with which labelled phosphate beomes incorporated into the nucleotide (Kluge *et al.*, 1970, working on *Yucca* and *Salix triandra* stems; and Gardner and Peel, 1972b, on bark strips from *Salix viminalis*). Also, it is not necessary to work with intact phloem, for Becker, Kluge and Ziegler (1971) and Gardner and Peel (1972b) have demonstrated the *in vitro* incorporation of ^{32}P into the ATP of sieve tube exudates. Further, the concentration of ADP is also high in phloem exudates (Gardner and Peel, 1972b, have measured ATP : ADP ratios of between 2 : 1 and 5 : 1 in stylet sap from willow), an observation which, again, favours the idea that the ATP has a high turnover rate which could be associated with energy supply to an active translocation process.

A different role for ATP has been proposed by Kennecke *et al.* (1971). One of the key glycolytic enzymes they detected was phosphofructokinase. This enzyme is inhibited by high ATP concentrations but is activated by high AMP concentrations, the latter compound stopping fructose-1,6-diphosphatase competing with phosphofructokinase. AMP concentrations in phloem sap are low (Kluge and Ziegler, 1964), so the inhibition of phosphofructokinase by ATP will not be relieved by AMP. Kennecke and his colleagues propose that the ATP in sieve tubes may be present to control the activity of the phosphofructokinase, although they have not as yet elaborated their thesis.

At the present time, we can only say that both the enzymic and energy 'potentials' of the sieve element appear to be high. Further work must be carried out, however, before we can unequivocally state that this metabolic potential is necessary for the longitudinal transport of solutes in sieve tubes.

The effect of localised application of low temperatures, metabolic inhibitors and anoxia on transport

Before proceeding to a detailed account of experiments using localised applications of metabolic depressors, it is most necessary to amplify the comment made in the introduction to this chapter concerning the interpretation of the data obtained from these experiments. Taking low-temperature treatment, there are two points which should be borne clearly in mind: (1) inhibition of transport does not necessarily mean that the mechanism is an active process; and (2) it cannot be justifiably argued that lack of inhibition implies the functioning of a passive rather than an active mechanism.

These statements may seem to be so sweeping that some further explanation is necessary. With regard to (1), there are at least two reasons which can be given in justification. As we have seen in the previous chapter (*Figure 7.1*, p. 121), it is highly probable that some energy, albeit rather a small proportion of the total, is required to maintain the structural integrity of the path; clearly, therefore, a reduction or complete cessation of this energy application could halt the functioning of a passive mechanism. Zimmermann (1958), for example, has obtained evidence that metabolic energy is necessary for the maintenance of sieve element permeability; thus, any interference with this could have drastic effects upon a pressure flow system. There is also the matter of viscosity changes in the flowing solution, brought about by alteration of temperature. As Duloy, Mercer and Rathgeber (1962) have pointed out, the viscosity of a 20% sucrose solution is 1.5 cP at 30°C, but at 5°C this rises to 3.15 cP. If pressure flow occurs in sieve tubes and if this obeys the Poiseuille formula, then the flow rate should vary inversely with the viscosity, i.e. a 50% reduction in movement should be induced by a fall in temperature from 30 to 5°C. Of course, the situation is more complex than this, for an increase in resistance consequent upon a rise in viscosity would increase the hydrostatic pressure gradient over the cooled portion of path; therefore some degree of compensation would be expected, leading to a partial recovery in translocation rate.

Statement (2) has been made because the majority of investigators have tacitly assumed that the treatments applied to the translocation pathway have achieved the objective of reducing or completely stopping metabolic energy turnover. As we shall see later, certain pieces of evidence have emerged recently which indicate that some metabolic processes can proceed, possibly at a sufficient level to maintain the working of an active mechanism even at low temperatures.

Generally, difficulties arise in the interpretation of inhibitor experiments rather similar to those catalogued above for low-temperature treatments. However, the former also raise certain problems peculiar to the application of exogenous substances to the phloem, one of the most worrying of which is the difficulty of ensuring that the inhibitor actually reaches the sieve elements. Even if this is accomplished, it is difficult to guarantee a sufficient concentration of inhibitor within the phloem to affect metabolic processes. A further problem with petiole- or stem-applied inhibitors is the possibility that any inhibition of transport might be brought about indirectly, owing to movement into the xylem and thence into the leaves, where loading processes could be affected. It is therefore necessary, with inhibitor experiments, to preclude any possibility of a reduction in sieve tube loading being the cause of an over-all decrease in transport rates.

Low-temperature experiments

Despite the criticisms which can be made against any conclusions drawn from experiments employing localised low-temperature treatments, the situation might hold a gleam of hope if only some consensus of opinion could be reached as to the actual effect of temperature. Sadly, however, no complete unanimity of view exists, investigators being divided into two camps: those in favour of inhibition by low temperatures (the majority) and those who believe low temperatures either have no effect or actually enhance the rate of movement.

One of the earliest workers on localised low temperature was Curtis, who in 1929 reported that cooling the petioles of *Phaseolus* to between 1 and 4°C produced a marked inhibition of translocation as measured by the loss in the weight of the lamina during the experimental period. Further experiments by Curtis and Herty (1936) confirmed these results by demonstrating that translocation was markedly reduced at 7°C as compared with that at 24°C.

Two of the most comprehensive investigations in recent years have been those performed by Webb and Gorham (1965b) and Webb (1967), using *Cucurbita*.

In the first of these publications temperature treatment was localised to the primary node and basal end of the primary petiole of 13-day-old plants. Labelled carbon dioxide was supplied to the primary leaf blade; therefore the labelled translocate movement to both the upper and lower parts of the plant could be influenced by the temperature treatment. Several hours were allowed for assimilation and translocation of the label before the various parts of the plant were assayed for activity. Webb and Gorham also took the precaution of measuring the transpiration and assimilation rates of the $^{14}CO_2$-fed leaf to ensure that there were no detrimental effects of the temperature treatment on these processes.

Their results may be summarised by reference to *Figure 8.1*. Varying the node temperature from 0 to 45°C did not produce any alteration in the rate of either transpiration or $^{14}CO_2$ assimilation.

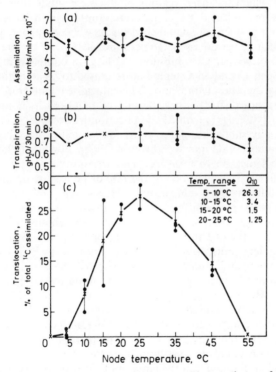

Figure 8.1 Effect of primary node temperature on: (a) assimilation of $^{14}CO_2$ by the primary leaf blade; (b) transpiration from the primary leaf blade; (c) translocation of ^{14}C through the temperature-treated node. Q_{10} = temperature coefficient of a process. (From Webb and Gorham, 1965b, courtesy of the National Research Council of Canada)

In complete contrast, [14C]sugar export was markedly affected by the node temperature, there being no significant movement through the treated node at 0°C. As the temperature was increased above 0°C, the rate of transport rose, reaching a maximum at 25°C, and then fell again to zero at 55°C. The inhibitory effect of low temperatures was reversible, normal rates of translocation being reached approximately 1 h after raising the temperature from 0 to 25°C. However, the effect of high temperatures could not be reversed, which showed that permanent damage had been caused to the system. Webb and Gorham (1965b) concluded that their data showed sugar transport through the node to be under direct physiological control, although they did not consider the process to be mediated by a streaming mechanism.

Essentially similar patterns were found by Webb (1967) in experiments in which localised regions of the stem, node, hypocotyl or petiole were subjected to temperature control. Both basipetal and acropetal movement of 14C-labelled sugars were completely inhibited over a 45 min period by a temperature of 0°C. A temperature of 10°C produced partial inhibition; 55°C gave complete inhibition; and between 13 and 35°C temperature ceased to be a limiting factor governing the rate of transport. These localised temperature treatments did not disturb either the rate of $14CO_2$ assimilation or the export of [14C]sugars from the application leaf blade.

Essentially similar results to those just described have been given by Mortimer (1961) and by Thrower (1965). Bowling (1968b) used an indirect method to measure sugar transport from the leaves of sunflower by determinations of the rate of potassium uptake by the roots at different stem temperatures. He showed transport to have a high temperature coefficient of approximately 3 in the range 0 to 25°C, although complete inhibition was not obtained at 0°C, as evidenced by the distribution of foliar-applied [14C]sugars.

In complete contrast, it is possible to quote work by investigators such as Went and Hull (1949), who showed the rate of exudation from detopped tomato plants to depend upon the sugar content of the roots. By applying a 7% sucrose solution to leaves 25 cm up the stem, the rate of exudation could be increased after 12 h. Thus, by varying the temperature of the stem along a 20 cm length between the leaves and the roots, the effect of temperature upon the rate of translocation could be measured in terms of the rate of bleeding. Went and Hull found that the amount of sugar transported gradually increased as the temperature was reduced; that is, transport showed a Q_{10} less than 1.

Hull (1952) also employed bleeding rate measurements on tomato, as well as [14C]sugar application to the leaves of this species and sugar beet, to follow temperature effects upon transport. Autoradiographic analysis of plants, subjected to a collar maintained at 1–3°C around either the petiole or the stem, showed equal or greater transport at low as compared with normal temperatures. Bohning, Kendall and Linck (1953) showed that stem elongation rates of tomato plants kept in darkness were directly proportional to the concentration of sucrose applied to a leaf blade over the range 0.0–0.04M. By varying the temperature of the stem along a short length below the sucrose-fed leaf, Bohning and his collaborators were able to measure effects upon rates of stem elongation and thus upon sugar transport from the leaf. They showed 24°C to be an optimum for transport, with a low value of the Q_{10} between 12 and 24°C of 1.5.

Using a different type of approach, Tammes, Vonk and van Die (1969) determined the effect of temperature upon phloem exudation from *Yucca* inflorescence stalks which were still attached to the plant. Cooling the tip of a stalk in melting ice slowed the rate of exudation, although the total dry matter was enhanced (*Figure 8.2*), as was the length of the exudation period. Tammes *et al.* thus concluded that low temperature had no marked effect upon longitudinal transport, although lateral movement into the sieve elements was very sensitive, since cooling the whole inflorescence top caused exudation to cease within 1 h.

What is particularly striking about the work described so far is that the effect of temperature seems to depend to some extent upon the species of plant employed. Whereas transport ceases at 0°C in *Cucurbita* (Webb and Gorham, 1965b) and in soyabean (Thrower, 1965), it does not do so in sunflower (Bowling, 1968b) or in willow at −1.5°C (Ford and Peel, 1966). Indeed, in the latter species transport of ¹⁴C-labelled assimilates does not cease completely in 3–6-year-old stems until a portion of the stem is cooled to −4°C (Weatherley and Watson, 1969), at which temperature irreversible damage is caused to the phloem cells by freezing. Thus, the possibility exists that one of the causes of variability could be species difference.

Two other factors which are probably also responsible for the variation in results have been indicated by the work of Ford and Peel (1967a) and Gardner and Peel (1972a), these being the stage of differentiation of the plant organ employed and the method used to measure the rate of translocation. In the first of these papers Ford and Peel set out to determine the effect of low temperature upon movement of ¹⁴C-labelled sugars in both young shoots and 3–4-year-old mature stems of willow. A schematic representation of their system is shown in *Figure 8.3*, the relative rate of translocation through portions of the phloem, maintained at normal and low

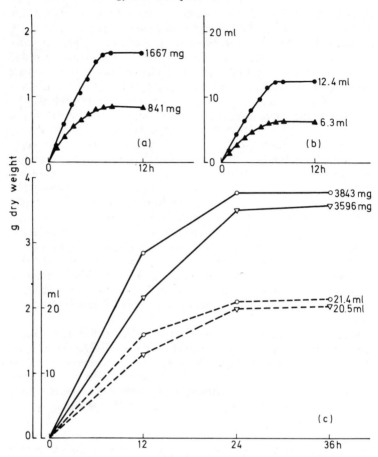

Figure 8.2 Exudate production and wound sealing at 20–25°C and 0°C, in the inflorescence stalk of one Yucca *plant: (a) and (b) represent the time course of exuded solutes and exuded volume at normal temperature in two series of measurements; (c) represents these also in two series of measurements, when the proximal 10–15 cm of the stalk was kept at 0°C. (From Tammes, Vonk and van Die, 1969, courtesy of the North Holland Publishing Co.)*

temperature, being measured by the rate at which ^{14}C-activity arrived in the honeydew from a colony of aphids sited below the experimental section of stem. The effect of low temperatures on [14C]sugar transport, relative to movement at 20–25°C, as found by Ford and Peel, is shown in *Figure 8.4*.

The most surprising feature of these data is that here, in a single species, we have results typical of both views which have been

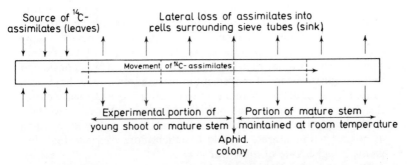

Figure 8.3 *Schematic representation of sink–conduit–source relationship in willow translocation systems. (From Ford and Peel, 1967a)*

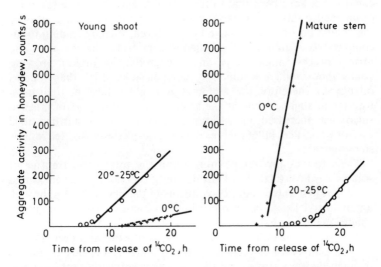

Figure 8.4 *Effect of cooling a portion of a young shoot or mature stem of willow to 0°C on the relative mass transfer of* [14]*C-labelled assimilates from the leaves to the site of an aphid colony. (From Ford and Peel, 1967a, courtesy of The Clarendon Press)*

described so far as to the effects of low temperatures on transport; in young shoots a reduction in transport was observed, while in mature stems movement was enhanced. Ford and Peel (1967a) rejected the idea that their data showed a fundamental difference in the mechanism of transport in the same organ at two different stages of differentiation. In explanation of their results, they invoked measurements of the radius of the sieve tubes: in young shoots this was found to be 12.6 μm; in mature stems, 19.0 μm. Thus, assuming a Münchtype pressure flow, a rise in the viscosity of the flowing solution, consequent upon a fall in temperature, would have a much more marked effect upon movement through the sieve tubes of young shoots than those of mature stems, the resistance of the former being nearly 16 times that of the latter. Further, it was considered that the enhancement of movement in mature stems could have been caused by a reduction of radial movement of labelled sugars out of the sieve tubes in the cooled portion. Thus, although the sieve tube stream might have been moving very much more slowly at low than at normal temperatures, it could be much more concentrated, thus giving a higher rate of mass transfer.

Of course, what Ford and Peel were measuring was not total sugar mass transfer but merely relative mass transfer of ^{14}C-label. Thus, could it not have been that in the mature stems the active secretion of unlabelled sugars from storage parenchyma into the sieve tubes of the cooled portion might have been reduced, thereby leading to an enhanced contribution of labelled sugars from the leaves? (The former process might be of small magnitude in rapidly growing young shoots, which would tend to act as sinks rather than sources of sugars.) Therefore, low temperature might reduce the total mass transfer of sugars, while at the same time the transport of labelled sugars was increased, i.e. a situation very similar to that described by Ford and Peel (1966) with labelled phosphate in isolated willow stem segments.

The effect of low temperature upon the total mass transfer of solutes was investigated by Gardner and Peel (1972a), who measured the flux rate of three major constituents of the sieve tube sap through an aphid stylet sited below a portion of a mature willow stem, under conditions of normal and low temperature (*Table 8.3*).

Their data give no support to the thesis of Ford and Peel (1967a), for low temperatures certainly did not enhance the flux of solutes through the sieve tubes, and the sugar concentration in the exudate always fell after low-temperature application. It could be concluded, therefore, that the form of the data extracted from localised low-temperature experiments may depend not only upon the species used in the investigation, and possibly upon the stage of its differentiation, but also upon the method used to measure rates of transport, pulse labelling being of doubtful efficacy in this connection.

A fourth factor, which seems very likely to play a prominent role in shaping the pattern of results from low-temperature experiments, is the phenomenon of recovery. In a short though excellent review of the effects of cooling various parts of the transport system, Geiger (1969) has indicated that two processes may be playing a role in recovery phenomena, viz. a short-term and a long-term recovery process.

Table 8.3 EFFECT OF COOLING A PORTION OF THE STEM OF A LEAFY WILLOW CUTTING TO 0°C FOR A PERIOD OF 24 h, ON THE CONCENTRATIONS AND FLUXES OF SUCROSE, ATP AND POTASSIUM IN SIEVE TUBE EXUDATE OBTAINED VIA APHID STYLETS. (FROM GARDNER AND PEEL, 1972a, courtesy *Springer-Verlag*)

Experiment		Volume flow rate, μl/h	conc., μg/μl			Flux, μg/h		
			Sucrose	ATP	K	Sucrose	ATP	K
1	A	1.339	90	0.71	7.5	120	0.95	10.0
	B	1.651	73	0.49	5.1	120	0.80	8.4
2	A	1.695	81	1.22	6.7	137	2.07	11.4
	B	0.472	78	0.67	11.3	37	0.32	5.3
3	A	1.537	80	1.30	2.7	123	2.00	4.2
	B	1.340	50	1.55	4.4	67	1.76	5.0
4	A	2.072	143	0.31	4.4	148	0.34	4.6
	B	2.174	89	0.31	3.9	97	0.35	4.2
5	A	0.860	116	0.53	7.1	50	0.23	3.1
	B	1.529	73	0.39	3.1	56	0.30	2.4

A = before cooling stem; B = after cooling stem.

Short-term recovery can be illustrated by the data of Swanson and Geiger (1967) on sugar beet. Cooling a 2 cm portion of the petiole to 1°C produced a rapid inhibition of transport, but within 2 h complete recovery took place (*Figure 8.5*). This type of recovery from cooling is, according to Geiger (1969), most likely to be due to an adjustment of inhibition, when the latter was caused by a mechanism such as an increase in viscosity. Thus, short-term recovery could be held to support the passive hypotheses.

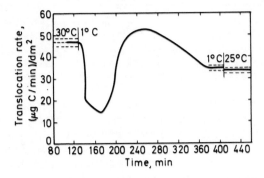

Figure 8.5 Time course of translocation through a 2 cm zone of sugar beet petiole at 1°C. Petiole zone temperature indicated above the curve. (From Swanson and Geiger, 1967, courtesy of the American Society of Plant Physiologists)

On the other hand, there are also examples of long-term recovery or acclimatisation. Geiger (1969) quotes work by Curtis and Herty (1936) and by Swanson and Böhning (1951) as examples of this phenomenon. In the latter work with bean, some recovery was observed between 65 and 135 h after localised path cooling to 10°C. Geiger (1969) sees long-term recovery as evidence for '. . . a mechanism for low temperature inhibition of translocation, in addition to direct inhibition of the key process which moves the translocate'.

It has already been pointed out that species difference could be a cause of the variability in data obtained by different groups of workers. Geiger (1969) gives a vivid illustration from work by Bayer of a quite different response to cooling, not between species but between two ecotypes of Canada thistle. Cooling to 0.5°C produced inhibition in both types, but with the northern ecotype rapid recovery occurred (*Figure 8.6*).

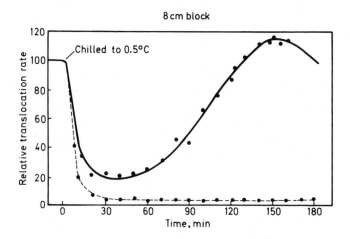

Figure 8.6 Comparison of recovery from chilling an 8 cm petiole zone to 0.5°C in the northern (solid line) and southern (dashed line) ecotypes of Canada thistle. (From Bayer, quoted in Geiger, 1969, courtesy of the Ohio Journal of Science)

It must now be clear to the reader that although experiments to investigate temperature effects upon transport are technically simple, the data extracted from them are very difficult to interpret and are subject to the effects of a large number of variables. Recovery is a process which, clearly, is of the utmost importance, for this may markedly influence the pattern of results obtained and, hence, any inferences as to the effects of path chilling upon transport. Much more work is needed before we can unequivocally decide exactly what low temperature does to the transport system. We certainly cannot conclude from experiments showing low-temperature inhibition that transport is an active process; apart from viscosity changes, cooling might produce increased path impedance due to callose formation, which would probably retard either a passive or an active system.

A further point which must be clarified is the effect of low temperature on phloem metabolism; for if it were shown that respiratory processes were still functioning after cooling, then the very keystone of low-temperature experiments would be removed. Active mechanisms could function at 0°C if energy were available. Weatherley and Watson (1969) concluded that some respiration was taking place at temperatures below 0°C (about 5% of the value at 25°C) in isolated pieces of willow bark. More direct evidence for respiration at low temperatures has been given by Coulson and Peel (1971). Using a technique similar to that described previously (Coulson and Peel, 1968; and p. 153), they were able to compare the relative rates at which ^{14}C-labelled sugars were respired in the phloem of willow stems at 25 and at 0.5°C. In both young shoots and mature stems some respiration of the translocated sugars took place at 0.5°C; indeed, in the mature stems the relative rate of breakdown at 0.5°C was greater than at 25°C. The detailed implications of this work are not clear, but at all events these results throw doubt upon the seemingly tacit assumption of many workers that at around 0°C respiration rates are completely insignificant.

Inhibitor experiments

The reader has probably decided already that inhibitor studies seem unlikely to lead to a quick solution of the problem of the transport mechanism, and this conclusion is certainly strengthened by the available data. Early experiments (Curtis, 1929, using atmospheres of nitrogen; Mason and Phillis, 1936, employing Plasticine) produced some indications that low oxygen tensions along the pathway may impede transport.

However, more recent work has cast doubt upon the validity of these data; for although there is good evidence that conditions of anoxia at the sink end of the system reduce the rate of transport (Geiger and Christy, 1971), the results of Willenbrink's work (1957) contradict those of the earlier investigators. Willenbrink dissected the tissues from around the central bundle of *Pelargonium* petioles and then surrounded the exposed area with a chamber, from which the air could be displaced by purified nitrogen. Phloem transport was monitored by taking leaf samples before and after experimentation for phosphorus and nitrogen analyses, or by foliar-applied fluorescein. No effect of localised anoxia was found in any instance, even when the experimental period was as long as 24 h.

Most of the work with metabolic inhibitors has, in contrast, demonstrated retardation of movement when these substances (cyanide, dinitrophenol, azide, etc.) were applied to the phloem, a notable exception being the work of Vernon and Aronoff (1952), using cyanide on soyabeans. Kendall (1955) studied the effects of a number of substances on the movement of [³²P]phosphates out of bean leaves by injecting them into the central cavity, dinitrophenol and fluoride producing a marked inhibition of movement. Willenbrink (1957), using the *Pelargonium* petiole system described previously, showed a reversible inhibition of transport by HCN gas. Other inhibitors, such as dinitrophenol, arsenite and azide, caused irreversible inhibition.

Both Kendall and Willenbrink were very conscious of the possibility that phloem transport might have been stopped owing to the movement of inhibitor up in the transpiration stream to loading sites in the leaf mesophyll, although neither produced any quantitative evidence to refute this possibility. Harel and Reinhold (1966) specifically designed their experiments to distinguish between the effects of dinitrophenol on labelled sucrose transport in the sieve tubes, as opposed to effects upon the loading processes in soyabean seedlings. When dinitrophenol was applied to a leaf, before or during [¹⁴C] sucrose application, transport of the label was markedly reduced. A similar inhibition was also found when the inhibitor was applied to the path with the [¹⁴C]sucrose application leaflet still attached to the plant. When, however, dinitrophenol was applied via the cut petioles of the primary leaves after removal of the ¹⁴C-application leaflet of the first trifoliate leaf, an apparent stimulation of [¹⁴C] sucrose transport was observed. These results led Harel and Reinhold to conclude that the inhibitory effect of dinitrophenol on transport found by earlier investigators was primarily due to suppression of solute-loading mechanisms in the leaf.

The site of inhibitor action certainly has not yet been clarified. Experiments such as those by Peel and Weatherley (1963) and Gardner and Peel (1972b), using aphid stylets, do not aid in the resolution of the problem, for in these experiments inhibitors were applied to isolated stem segments or to the cambial surface of bark strips of willow. In these cases all the inhibitors (dinitrophenol, oligomycin and fluoride) caused cessation of stylet exudation, which in certain instances was partially reversible by replacement of the inhibitor with water or a solution of ATP. Although Gardner and

Peel came to the conclusion that ATP from oxidative phosphory-lation is ultimately necessary for transport, energy may only be utilised to any degree in the sieve tube loading processes. It also appeared from their work that the ATP present in sieve tube exu-dates may not be directly associated with transport, since stylet exudation could continue for 1–2 h after the concentration of ATP in the sap had been reduced to non-detectable levels by fluoride.

A recent, very comprehensive, piece of work on the site of cyanide inhibition of transport from sugar beet leaves has been reported by Ho and Mortimer (1971). Cyanide at a concentration of 0.5M was applied to the petiole of a leaf, 12 cm from the base of the lamina, entry of the inhibitor into the phloem being facilitated by removal of some of the extravascular tissues; then the blade was exposed to $^{14}CO_2$ 3 min after cyanide application. The initial experiments showed that cyanide treatment caused more of the ^{14}C-labelled assimilate to be retained in the leaf blade and less to be contained within the conducting tissues of the petiole when compared with water controls (*Table 8.4*).

However, Ho and Mortimer then went further than previous workers. Having demonstrated, by the use of ^{14}C-labelled cyanide, that the inhibitor rapidly accumulated in the lamina, they proceeded to examine any possible effects upon photosynthesis and vein-loading mechanisms. The labelled products of photosynthesis and their bio-chemical interactions were investigated in cyanide-treated and control leaves. Under their experimental conditions the ^{14}C-uptake by the cyanide leaf was comparable to that of the control, although the amount of ^{14}C-label in the former leaf was considerably greater

Table 8.4 EFFECT OF KCN PRETREATMENT ON DISTRIBUTION OF ASSIMILATED $^{14}CO_2$ AFTER 10, 20, 30 AND 50 min (KCN APPLIED TO PETIOLE 3 min BEFORE $^{14}CO_2$). (FROM HO AND MORTIMER, 1971, courtesy *National Research Council of Canada*)

Time, min		Total ^{14}C, %		Exported ^{14}C, % (conducting tissue)		
		Green tissue	Conducting tissue	1st-order veins	Midrib	Petiole
10	control	96.7	3.3	68.8	30.0	1.2
	KCN	98.4	1.6	82.0	17.9	0.1
20	control	91.8	8.2	28.4	48.9	22.6
	KCN	97.7	2.3	60.3	39.2	0.5
30	control	85.8	14.2	16.8	38.8	44.4
	KCN	96.0	4.0	59.2	40.2	0.6
50	control	81.2	18.8	16.1	25.3	58.6
	KCN	96.7	4.3	60.8	38.3	0.8

after 50 min from the start of assimilation than in the latter. The labelling rate of sucrose was, however, similar in both instances. Thus, they concluded that cyanide had no detectable effect upon the rate of photosynthesis and the subsequent formation of its phloem-mobile product sucrose.

Further experiments concerned with possible effects of cyanide upon vein loading were carried out using autoradiography of treated and control leaf blades. Clear indications were obtained that the inhibitor had no detectable effect upon loading, for the ^{14}C-translocate moved into the minor veins of the treated leaf faster than into those of the control; although loading was unaffected by cyanide, subsequent movement in the phloem was, for the minor veins of the cyanide-treated leaf were still heavily labelled 50 min after $^{14}CO_2$ application, whereas the image of those in the control leaf was fading after this period of time.

Phloem metabolism and its relationship to transport is undoubtedly a confusing field, particularly for those attempting for the first time to gain an insight into the subject. What has been attempted here is a review of our current knowledge and an indication of the difficulties involved in the interpretation of the data. To conclude this chapter, it might be helpful briefly to summarise the main points which have emerged:

(1) Phloem undoubtedly has a high metabolic potential when compared with parenchyma tissues, but we do not know how much of this potential, if any, is necessary for longitudinal movement.

(2) Localised low-temperature application to the phloem pathway results in a retardation of transport, although this effect is transitory in many plants. It is not possible to conclude from these effects whether transport is mediated via an active or a passive mechanism, since seemingly plausible explanations can be conceived which could accommodate either type of mechanism.

(3) It appears very probable that longitudinal phloem transport can be reduced or stopped by metabolic inhibitors. However, this does not necessarily rule out the operation of a passive mechanism, for this might require some energy to maintain the integrity of certain structures within the sieve elements, e.g. the semipermeability of the plasmalemma is a prerequisite for pressure flow, or a particular configuration of filaments might be essential for surface movement, both of which could be affected by a dearth of metabolic energy.

9

Simultaneous bidirectional movement and the kinetics of multiple tracer experiments

It is implicit in certain of the hypotheses described in Chapter 7 (pressure flow and electro-osmosis) that all solutes in one sieve tube should travel in the same direction and at very similar velocities. Thus, any study which provides a demonstration of simultaneous bidirectional movement must give support to some type of streaming or multimodal transport process. As with most facets of phloem physiology, however, theory is a good deal easier than practice. While it is technically feasible to execute experiments on simultaneous bidirectional movement and its associated processes, the data obtained can be interpreted in many ways, which leads to a situation which is even more confused, and confusing, than that described in Chapter 8.

The root cause of our difficulties resides, as always, in the complexity of the phloem tissue, and there seems little hope of resolving the technical problems associated with these studies in the near future. Indeed, so daunting is the problem of extracting meaningful data from multiple-solute transport studies that certain investigators now consider them to be virtually useless; Canny (1971), for instance, has described them as 'will-o'-the-wisp' experiments.

Although one has a certain sympathy with this view, the fact remains that the key to progress is experiment, and we cannot deride those who have attempted to clarify a problem in this way, for efforts in the past are merely steps on the way to achieving the final goal. Thus, in the present chapter it is intended to describe work to date, without attempting to draw any irrevocable inferences, but at the same time to constructively criticise the form of the experiments in the hope that this may eventually lead to some progress.

Simultaneous bidirectional movement

Broadly speaking, there are two types of experiments, both of which are in all essentials the same, which have been used to investigate this problem: (1) manipulation of the source–sink relationships of two solutes already present within the plant, or application of two tracers (frequently radioactively labelled) to spatially separated points on the transport path; and (2) following the movement of certain substances (viruses and herbicides) in relation to the presumed direction of the assimilate flow. Strange though it may seem, investigators who have employed method (1) have almost all come to the conclusion that bidirectional movement can occur, while those using method (2) have, with few exceptions, reached the opposite conclusion!

Studies on the movements of two solutes

Two of the earlier studies were carried out on cotton by Mason *et al.* (1936) and Phillis and Mason (1936), who investigated the movement of carbohydrates and nitrogen in opposing directions.

Mason and his co-workers showed that most of the mineral nitrogen absorbed by the roots moved upwards in the xylem, with the phloem taking but little part in its transport. However, when plants grown in culture solution were deprived of their nutrient supply, nitrogen was exported from the leaves of the basal region to the upper parts of the plant. Moreover, since this export was completely prevented by ringing the stem above and below the basal region, it appeared that the mobilised nitrogen moved up in the phloem. Mason *et al.* also demonstrated that carbohydrates moved down from the apical to the basal region during the experimental period when the mobilised nitrogen was moving upwards. Although they concluded that their data could provide support for simultaneous bidirectional movement, they did not believe that a firm conclusion on this issue was justified.

However, Phillis and Mason reached the firm inference that bidirectional movement was possible. In their experiments cotton plants were divided into two groups, the main axis of the plants in each group being divided into an apical and a basal portion; both groups were ringed below the basal region and a ring was also interposed between the two regions in one group. Phillis and Mason then proceeded to alter the source–sink relationships of the two nutrients, carbohydrate and nitrogen, by maintaining the leaves of the apical portion in the light and covering the leaves of the basal portion in paper bags. After 14 days the plants were analysed for nitrogen and carbohydrate content, these analyses revealing that where no ring had been made between the apical and basal portions of the stem, nitrogenous materials had moved upwards and carbohydrates downwards.

These results do not necessarily mean, however, that true simultaneous bidirectional movement (*Figure 7.2a*, p. 123) had taken place in the cotton plants. In both investigations the experimental period was of considerable duration (14–24 days). Therefore, it is quite possible to imagine unidirectional movement occurring at different times in a sort of see-saw action. Even if movement of the two nutrients was simultaneous, there is no evidence from these experiments that it took place in the same sieve tubes; different sieve tubes or even different parts of the phloem could have been responsible.

Similar criticisms can be made of the work of Turner (1960). Using *Pelargonium* plants, ringed in the lower stem region, this investigator showed the accumulation of nitrogenous compounds above the ring to be markedly increased by removal of the stem apices; sucrose accumulation, on the other hand, was but little affected. Turner argued that his results indicated an independence of nitrogen and carbohydrate transport, even though his experiments were of considerable duration (6 days) and he provided no evidence as to whether both solutes had moved within the same sieve tubes.

A rather more cogent piece of evidence in favour of simultaneous bidirectional movement was produced by Palmquist in 1938. Using *Phaseolus*, Palmquist demonstrated that fluorescein moved in the phloem of well-watered plants when applied in solution to a terminal leaflet, since under these conditions it would not pass a scalded portion of the petiole. By keeping plants in darkness for 15 h, Palmquist demonstrated a movement of carbohydrate out of the lateral leaflets and a concomitant movement of fluorescein into them, when the latter compound was applied to the terminal leaflet. He also demonstrated that carbohydrate could move into starved leaflets, as evidenced by changes in weight, while fluorescein could move out.

Palmquist argued that since the petiolule of a bean leaflet only contains one vascular bundle, the two compounds must have moved in opposite directions in the same sieve tubes. However, although Palmquist's experiments were of relatively short duration, it is clear that his work did not satisfy the stringent conditions necessary for an unequivocal demonstration of simultaneous bidirectional movement. There is no evidence from his experiments that movement in the sieve tubes did not reverse; also, fluorescein is not a naturally occurring solute and so it suffers from the criticism that it could move in the sieve tubes by a different mechanism from that of carbohydrate.

Some early work on simultaneous bidirectional movement using radioactively labelled solutes was carried out by Chen in 1951. Using the stripping technique of Stout and Hoagland (1939, Figure 1.6a), in order to preclude lateral movement of tracer from the xylem to the stripped bark of *Pelargonium* plants, Chen applied $^{14}CO_2$ to a leaf above the stripped portion and [^{32}P]phosphate to a leaf below or vice versa. After a period of 12–17 h from tracer application, both isotopes were detected in the stripped portion of bark. Thus, one had moved upwards while the other had travelled downwards. However, Chen's work did not take into account the possibility of the tracer which was applied to the lower leaves being moved up in the transpiration stream, with subsequent export from the upper leaves down the phloem.

Two most interesting studies having considerable relevance to the present discussion have been reported by Biddulph and Cory (1960, 1965), in which the movement of radioactive tracers was followed in red kidney bean plants by autoradiography.

In the first publication $^{14}CO_2$ was administered to the lowermost trifoliate leaf, followed 5 min later by [^{32}P]phosphate to the second trifoliate leaf. After a 30 min migration period, a piece of the internode between the two application leaves was removed, the phloem tissues being separated from the xylem. Autoradiograms were prepared showing the distribution of both tracers together, and of each individually. Scanning of the autoradiograms with a microdensitometer then revealed the distribution of each tracer in the various phloem bundles.

The most striking result of this work was that two apparently quite different types of bidirectional movement were detected. In the first, upward and downward movement of the tracers occurred in separate phloem bundles, this type of transport seeming to occur mainly in the older, more mature, parts of the phloem. In the second type bidirectional transport of the two tracers took place in the same strand of young phloem cells lying near to the cambium. The observed patterns of movement seemed so clear-cut that they forced Biddulph and Cory to conclude that two quite different mechanisms might be operating—a pressure flow type in the older sieve elements and a protoplasmic streaming type in the young parts of the phloem.

Certainly, it seemed that these results could provide an admirable way out of the dilemma confronting physiologists, for they gave an explanation of the conflicting data obtained from other work, some of which supported a passive mechanism, some an active one. However, it now seems that the 1960 work of Biddulph and Cory was based upon an unsound technical basis, for in a later paper (1965) they performed a comprehensive study of the distribution patterns of ^{14}C-labelled metabolites in bean plants.

After having shown the lower leaves to export primarily to the root and the upper leaves to the apex, while intermediate leaves exported in both directions, Biddulph and Cory then proceeded to try and identify the precise phloem channels in which movement occurred, using a combination of autoradiographic and fluorescence techniques. They were unable to demonstrate any counter-movement to the main direction of sucrose movement in mature bundles. In 30 cases examined where an upward movement of labelled assimilate occurred past a downward flow of unlabelled sugar, these movements always took place in separate vascular traces. Biddulph and Cory went on to suggest that their earlier (1960) demonstration of bidirectional movement in the same bundle was probably due to an overhydration of the freeze-dried tissues, which allowed labelled sucrose to move into tissues in which it did not occur before sampling of the experimental plants. Despite their retraction on the issue of simultaneous bidirectional movement, Biddulph and Cory still felt that their 1965 data supported the concept of a bimodal transport system with export from mature leaves resembling a mass flow and that from younger leaves a 'metabolic' process. The experimental basis for these conclusions seems, however, to be remarkably tenuous.

Although it might appear that aphids, since they tap a single sieve element, could help to clarify the problem of bidirectional movement, reference to *Figure 7.2* (p. 123) shows that this is not necessarily the case; puncture of a sieve tube by an aphid stylet could well initiate flows in the direction shown in *Figure 7.2 (b)* or (*c*). There is certainly no doubt that if two tracers are applied either side of an aphid colony, then both can occur in honeydew excreted by the insects.

Eschrich (1967) used the aphid *Acyrthosiphon pisum* (Harris) in conjunction with plants of *Vicia*, employing the experimental arrangement shown in *Figure 9.1*, the tracers being ^{14}C-assimilates (supplied as NaH^{14}CO$_2$) and fluorescein. Individual drops of honeydew excreted by the aphids were collected on to a rotating disc of cellophane and then assayed for tracer distribution. Eschrich demonstrated that a large proportion of the honeydew drops contained both tracers. This, he concluded, was either the result of bidirectional movement within a single sieve tube or the operation of a 'homodromous loop path', i.e. a situation in all respects the same as that shown in *Figure 7.2 (c)*. Eschrich dismissed the possibility of flows similar to those depicted in *Figure 7.2 (b)* occurring, for he showed that the 'attractive force' of a single aphid was insufficient to change the direction of transport.

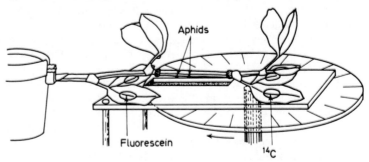

Figure 9.1 Experimental arrangement for the study of the simultaneous movement of fluorescein and [14]C-*assimilates in* Vicia faba *plants. (From Eschrich, 1967, courtesy of Springer-Verlag)*

Essentially similar, though slightly more sophisticated, experiments were performed by Ho and Peel (1969b) using [14]C- and ^{3}H-labelled sugars and [^{32}P]phosphates in willow stems, sieve tube sap being collected as honeydew from individuals of *Tuberolachnus salignus*. As with the work of Eschrich, a considerable number of aphids produced honeydew which contained two tracers by the termination of the experiment. In 'mature stem' experiments lasting from 24 to 48 h, six of the twelve aphids produced double-labelled drops, five single-labelled and one unlabelled. In 'young shoot' experiments of 24 h duration, nine of the ten aphids produced double-labelled drops, the other producing unlabelled honeydew. What may be of greater significance was that, of the twenty-two aphids employed, six produced double-labelled honeydew *from the first drop*.

It was this last observation which led Ho and Peel to infer that simultaneous bidirectional movement might be possible, although they pointed out that their technique did not preclude lateral movement of one tracer from one sieve tube into an adjacent one which was transporting the other tracer. Ho and Peel, however, confirmed Eschrich's observation that a single aphid is but a minor sink, for in one experiment an aphid feeding only 1 cm from the site of [^{14}C] sucrose application produced unlabelled honeydew over a period as great as 38 h.

What appears without any doubt to be the most cogent evidence in support of simultaneous bidirectional movement is the histo-autoradiographic study of [^{14}C] and [^{3}H]sugar movement in *Cucurbit* plants reported by Trip and Gorham (1968a). Tritiated glucose was introduced into 3-week-old plants through a flap cut from a side vein of a mature leaf for a period of 3 h, after which $^{14}CO_2$ was supplied for a further 20 min to an immature leaf of the same plant. The plant was then frozen and prepared for assays of the distribution of the two tracers.

Figure 9.2 shows the typical distribution of ^{3}H- and ^{14}C-labelled compounds in the plants employed. The gradient of ^{14}C-activity was in the reverse direction to that of tritium, which showed that bidirectional movement had occurred. Moreover, chromatography of tissue extracts revealed that the major proportions of both ^{3}H- and ^{14}C-activities were present in the form of the transport sugars stachyose, raffinose and sucrose, with only small quantities in the phloem-immobile hexoses, glucose and fructose.

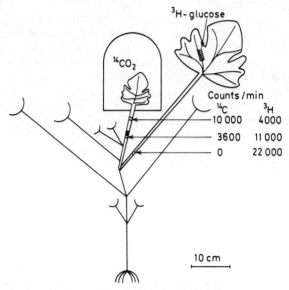

Figure 9.2 Diagram of a squash plant showing method of supply and gross distribution of the tracers, [^{3}H]glucose, supplied for 3 h, and $^{14}CO_2$ for 20 min. Autoradiographs were made from the blackened section of the petiole of the immature leaf. (From Trip and Gorham, 1968a, courtesy of the American Society of Plant Physiologists)

Further analyses of ^{3}H- and ^{14}C-distribution were performed by cutting serial sections of part of the petiole of the immature leaf to which the $^{14}CO_2$ had been supplied (*Figure 9.2*). Histoautoradiographs were prepared showing both ^{14}C and ^3H together and ^{14}C alone, 'streaks' on the autoradiographs being found which had been produced by radiation from both isotopes.

Having proceeded thus far, Trip and Gorham then discuss the validity of their technique in relation to the movement of both tracers within a single sieve tube. As they point out, it is possible for a streak on an autoradiograph of a 10 μm thick longitudinal section of 30 μm diameter sieve tubes to refer to superimposed portions of two adjoining sieve tubes, as well as to sections of the same sieve tube. Also, cytological details tend to be obscured in autoradiographs. Therefore, it is not possible, merely by viewing the section and film, to state categorically (a) whether a streak was produced by both tracers in the same sieve tube, or (b) whether the tracers were present in two separate sieve tubes. Trip and Gorham, however, argued that if the situation was as in (b), then grain counts due to the two labels would differ in ratio between the ends and middle of the streak.

Since they were unable to detect any difference in the ratio of ^3H to ^{14}C between transects across the ends and middle of the streak, they inferred that the upward-moving tritium and the downward-moving ^{14}C were present in the same sieve tubes. Their final conclusion was that the pressure flow hypothesis was inadequate to explain their observations and that '. . . while a mass flow of solution may account for movement in one direction, in some way the sieve tube protoplast must control and direct the translocation process to permit concurrent movement in opposite directions'. Trancellular streaming seemed to Trip and Gorham the most likely mechanism which would explain their data.

Transport of viruses, growth regulators and labelled assimilates into and out of mature leaves

It is not the purpose of this section to give a comprehensive account of the patterns of virus and herbicide movement in plants, since an excellent review of this topic has recently been provided by Crafts and Crips (1971), which incidentally includes information on the transport of a whole range of chemotherapeutants and pesticides. What has to be considered here is the evidence, particularly from studies with viruses and herbicides, which is pertinent to the problem of simultaneous bidirectional movement. To get to the core of the matter, this evidence involves one main issue, i.e. whether exogenous substances can, or cannot, move into mature leaves against the assimilate stream. If this is shown to be possible, then it can be argued that the phloem system must be capable of allowing simultaneous bidirectional movement.

Much of the work concerning the relationships between virus and assimilate movement has been done by Bennett. This worker (1940a) divided plant viruses into three main groups on the basis of their transport properties: (1) viruses limited to phloem which can move through the sieve tubes at velocities equal to those found for assimilates, (2) viruses limited in their distribution to parenchyma tissues in which they spread but slowly, and (3) certain types which can multiply in parenchyma, but which can subsequently invade phloem in which they are able to spread rapidly. Clearly, it is types (1) and (3) viruses, producing diseases such as vein and leaf distortion, leaf rolling and the mosaic infections, which are of greatest interest to phloem physiologists, since there is abundant evidence to show that their movement within the plant is closely correlated with assimilate transport.

Bennett (1940b), carried out a study of the movement of mosaic virus in tobacco. In plants with a horizontal stem and a basal sucker, basipetal movement of the virus in the main stem was rapid, while acropetal movement was slow. However, in plants that possessed considerable sink activity at the apical end in the form of maturing seeds, acropetal movement was shown to be very rapid. In vegetative plants acropetal movement could be accelerated by darkening, or by removal of the mature leaves. In darkness basipetal movement was very slow, a fact not unexpected if the virus followed the assimilate stream, for under these conditions the latter would be nonexistent.

In an earlier paper Bennett (1937) investigated the relationships between the movement of curly top virus and the presumed patterns of assimilate movement in triple-crowned plants of sugar beet. Inoculation of one crown led to the rapid transport of the virus into the roots, but movement into the other crowns was very slow, as evidenced by the fact that no symptoms had been produced even after 140 days from inoculation. Since curly top is a phloem-limited virus, its transport pattern was clearly following that of the normal phloem stream. Moreover, movement of the virus from the inoculated to the non-inoculated crowns could readily be achieved by procedures designed to produce a flow of assimilates into the latter; defoliation or darkening of the non-inoculated crowns caused a rapid onset of symptoms in these organs.

Anatomical observations on plant virus infections have been reviewed by Esau (1967); more recently, Esau, Cronshaw and Hoefert (1968) have performed investigations with the electron microscope on the relation of the beet yellows virus to the phloem and movement in the sieve tubes. In the minor veins of sugar beet the virus particles were present in both parenchyma cells and mature sieve elements. In the former they were usually confined to the cytoplasm; in the latter, scattered throughout the cell. Some of the sieve elements containing virus particles appeared normal, while others showed signs of degeneration. Virus particles were present in the sieve plate pores, the plasmodesmata connecting sieve elements with parenchyma cells, and the plasmodesmata between parenchyma cells. Esau and her colleagues reached the conclusion that the distribution of the virus in the phloem was compatible with the concept that the particles move through the sieve tubes by a passive pressure flow mechanism.

There are several examples in the literature of studies with growth regulators which have led workers to conclude that a close correlation exists between the transport patterns of these substances and those of endogenous sugars. To take two examples concerning the movement of 2,4-dichlorophenoxy (2,4-D) acetic acid, the work of Mitchell and Brown (1945) and Rohrbaugh and Rice (1949) can be cited. Both groups of workers employed bean plants, the growth regulator being foliar-applied, a measure of the extent of its transport being obtained from the degree of stem curvature produced during a given time from application.

Mitchell and Brown (1945) showed the growth regulator not to be readily transported out of leaves whose sugar content had been reduced by extended periods in darkness or CO_2-free air in light, i.e. from leaves which would not be expected to be transporting sugars. Also, 2,4-D was not translocated from sink regions such as young, growing leaves, which would themselves be importing foodstuffs.

Rohrbaugh and Rice (1949) extended these findings by demonstrating that 2,4-D was not transported out of destarched leaves unless sugars were also supplied (*Table 9.1*), i.e. it was presumed that these applied sugars were also translocated, carrying the growth regulator with them.

Stout (1945), working on the translocation of the reproductive stimulus in sugar beet, also came to the conclusion that the transport of this was closely associated with that of endogenous sugars. Separate shoots of annual beet plants were subjected to different photoperiodic regimes, viz. long days, short days or continuous darkness. Parts exposed to long days developed seed stalks, as did those kept in continuous darkness if these were connected in parts in long day regimes. Parts kept in short days did not produce reproductive structures, even though they were joined to long day parts

Table 9.1 EFFECT OF APPLICATION OF SUGARS ON STEM CURVATURE OF BEAN PLANTS TREATED WITH Na(2,4-D) IN THE DARK. (FROM ROHRBAUGH AND RICE, 1949, courtesy *University of Chicago Press*)

Group	Treatment	No. of plants	Stem curvature, °
	Ends of leaves in sucrose solution		
A	100 µg Na(2,4-D)	20	33.0
B	No Na(2,4-D)	20	0.5
	Ends of leaves in fructose solution		
A	100 µg Na(2,4-D)	20	46.5
B	No Na(2,4-D)	20	0.5
	Ends of leaves in glucose solution		
A	100 µg Na(2,4-D)	20	44.0
B	No Na(2,4-D)	20	3.2
	No sugar or 2,4-D applied	24	0.0

Stout explained these results by assuming a close association between reproductive stimulus and sugar transport; the parts in continuous darkness flowered because they received sugars and the flowering stimulus from long day parts, while the parts in short days synthesised sufficient carbohydrate to be independent of the long day portions.

It has been mentioned in Chapter 5 that mature leaves, maintained in the light, import little if any carbohydrates and other nutrients, and this observation has often been cited by those who believe in a unidirectional mechanism such as pressure flow to support their belief. However, there are several pieces of work which indicate that some activity can be detected in mature leaves when relatively high doses of labelled assimilates are given to other mature leaves of the same plant (Thaine *et al.* 1959; Canny and Askham, 1967).

The latter workers have attempted to use this observation as evidence for a streaming process in transcellular strands. According to Canny (1962) and Canny and Phillips (1963), such a mechanism would be a diffusion-analogue type, i.e. it would show many of the characteristics of a thermal diffusion process, except that it would proceed at a much greater rate. Thaine's concept of peristalsis in transcellular strands (1969) would not constitute a diffusion-analogue system, since this is effectively a pressure flow system. Canny and Askham (1967) argued that if movement is indeed analogous to diffusion, then the injection of tracer into the transport system should lead eventually to the spread of the label throughout the whole phloem (*Figure 9.3*), even into the mature leaves as far as the boundary at which sugars are loaded into the sieve elements. Thus, by their argument, if one mature leaf of a plant is supplied with labelled sugar, the veins but not the laminae of the other mature leaves should also, after a period of time, come to contain a label.

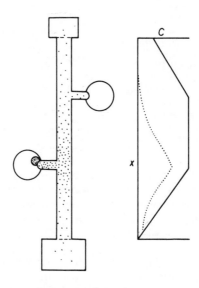

Figure 9.3 Schematic representation of the plant transport system (left). Sources (leaves) are represented by circles, sinks by rectangles. The concentration of tracer spreading through the system is shown by dot frequency (C) against distance (x) up and down the plant. Non-tracer sugar is shown as a continuous line, tracer as a dotted line. Non-tracer sugar is in a steady state, tracer sugar is not. (From Canny and Askham, 1967, courtesy of The Clarendon Press)

Canny and Askham attempted to support their thesis by experiment. Having infested plants of *Vicia* with *Aphis craccivora*, some of which settled on the veins of mature leaves, they applied $^{14}CO_2$ to a small portion of one leaflet of a mature leaf. After 6 h the plants were harvested and dried and autoradiographs were prepared. While the mature leaves produced, at the most, faint images on the film, the aphids feeding upon them were quite heavily labelled, which showed that some tracer had entered the phloem of the mature leaves on which the aphids had been feeding.

Since activity was clearly present within the sieve elements upon which the aphids had fed, Canny and Askham concluded that this could only have occurred if transport was mediated via a diffusion-analogue mechanism. They dismissed any possibility of this situation arising with a pressure flow process, for surely no activity at all should enter the leaf in the case of such a unidirectional mechanism unless the aphids themselves were acting as strong sinks, drawing activity in from more distant parts of the plant. This latter possibility was considered to be most unlikely, in view of the fact that the aphids were feeding on a mature, photosynthesising leaf where the sugar 'potential' should surely be high enough to supply the demands of the aphids without the necessity of drawing on solutes from other sources.

Before proceeding with a consideration of studies which aim to show that different solutes can move at different velocities, it might be helpful to analyse the situation with respect to simultaneous bidirectional movement. As a very general statement, it can be said that a considerable number of physiologists now concede that simultaneous bidirectional movement is possible. (Even Spanner in his 1970 publication seems to believe that it occurs and has attempted to accommodate the process into his unidirectional electro-osmotic hypothesis.)

The belief in simultaneous bidirectional movement is not, of course, based upon the results of long-term experiments in which no account can be taken of possible movement in opposite directions at different times or in different sieve tubes (Phillis and Mason, 1936; Palmquist, 1938; Chen, 1951) but upon the aphid experiments of Eschrich (1967) and particularly upon the results of Trip and Gorham (1968a). Shortcomings in the aphid work have already been indicated; a consideration of the autoradiographic study is, therefore, now appropriate.

Taking the assertions of Trip and Gorham that both the ^{3}H- and the ^{14}C-labelled sugars were present within the same sieve tube as being correct, there are several means which could have produced this situation other than the operation of simultaneous bidirectional movement. Possibly the most cogent argument against the conclusions of Trip and Gorham is based upon the fact that the leaf to which they applied the $^{14}CO_2$ was immature, and there is ample evidence in the literature (Jones *et al.* 1959; Jones and Eagles, 1962; Thrower, 1962) that immature leaves both import as well as export assimilates. It is not, therefore, unreasonable to argue that both tracers could have appeared within the same sieve tube as a result of lateral transfer between a sieve tube importing [^{3}H]sugars and one exporting [^{14}C]sugars. Another possibility is that an actual reversal of flow might have taken place within the sieve tubes during the 3 h feeding period of the tracers. Yet another possibility, albeit less likely, is that some of the [^{3}H]sugar could have migrated into the xylem and thence into the vascular tissue of the ^{14}C-fed leaf. The evidence

showing the transport system to be capable of supporting simultaneous bidirectional movement is good, but by no means conclusive.

Similarly, arguments in favour of a unidirectional mechanism
based upon patterns of virus and growth regulator movement do not
hold up when subjected to close scrutiny. Any type of transport
mechanism must be under some form of control and this is certainly
provided by sources and sinks (Chapter 5). Thus, it is difficult to
understand the repeated assertions of Crafts and Crisp (1971) that
control of transport by sources and sinks is indicative of the operation of a pressure flow mechanism. Certainly, both Spanner (1958)
and Canny (1962) accept that control of rate and direction reside
primarily in events at either end of the system.

Thus, it is not surprising that viruses and growth regulators move
in the same direction as the assimilate stream; even with a transcellular strand mechanism in which simultaneous bidirectional
movement is possible, in the presence of a powerful source–sink
system it is most likely that all the strands would be transporting in
the same direction. Such a situation would appear to be likely whatever mechanism was responsible for movement in strands, be it
streaming (Canny, 1962) or a mass flow produced by peristaltic
contractions (Thaine, 1969). Thus, movement of exotic materials in
a direction opposite to that of the main assimilate stream would be
minimal or non-existent, certainly insufficient to be detected by the
relatively insensitive methods used for monitoring the presence of
viruses or growth regulators in the work which has been described.

Although Canny and Askham (1967) state that they are '. . .
habitually suspicious of the *experimentum crucis* . . .' it is clear from
their paper that they believe that their work constitutes such a rare
phenomenon. Their arguments, however, are too facile to allow of
more than a passing belief in the veracity of their conclusions; while
the latter seem sound in the context of a simple physical system,
several objections can be raised to them based upon the complexity
of the whole plant.

Firstly, we do not yet completely understand how the movement
of assimilates into and out of leaves is controlled. Although solute
potential clearly plays a major role, no doubt other factors less
amenable to analysis such as hormone balance operate, and it may
be unjustified to conclude that the import of small quantities of
tracer into a mature leaf necessarily precludes the operation of a
unidirectional mechanism; a few sieve tubes could still import while
the majority exported sugar. Secondly, Canny and Askham do not
seem to consider the possibility that the radiocarbon appeared in
the veins of the aphid-infested leaf via the xylem. Small quantities
could certainly be transported in this way, in view of the ease with

which radial movement between the two transport systems is achieved.

The picture at present is admittedly not very clear, but work must be continued to solve the problem. It is almost certain that refinements of extant techniques and the use of entirely new ones will eventually allow us to state categorically whether simultaneous bidirectional movement does or does not exist.

Multiple solute transport in one direction

The great difficulties encountered in the accurate measurement of velocities by radiotracer techniques have been fully discussed in Chapter 4, and it is not necessary to enumerate them further. It is sufficient here to state that the majority of experimenters have not taken any great account of the effect of either differential rates of tracer removal from or secretion into the sieve tube; thus what has been measured is not velocity but relative mass transfer. Since it is quite possible for two or more solutes to move by pressure flow with different rates of mass transfer, the conclusions of many workers are therefore completely unjustified. These considerations particularly apply to the study of Vernon and Aronoff (1952), who demonstrated differential velocities of sucrose, glucose and fructose transport in soyabean stems. However, they did indicate that the cause of the differences might reside in the initial step from the mesophyll to the sieve elements. It is also very probable that the so-called 'preferential translocation' of different sugars in white ash and lilac (Trip *et al.*, 1965) was also, as recognised by these workers, largely due to different rates of removal from the sieve tubes.

A piece of work often quoted as demonstrating different velocities of transport is that of Swanson and Whitney (1953). One objective of their experiments (on *Phaseolus*) was to measure the effect of temperature on movement, but they also attempted to determine whether two members of a pair of foliar-applied radiotracers moved with different velocities relative to each other. Since they measured the relative velocities by the change in the ratio of the activities of the two isotopes of a pair with increasing distance from the application site, Swanson and Whitney avoided the problem caused by different rates of secretion of each tracer into the sieve elements of the leaf. What they could not do, however, with their techniques was to distinguish between moving and non-moving tracer in the stem segments they analysed. Thus, although some tracers apparently moved with different relative velocities (*Table 9.2*, ^{137}Cs and ^{32}P), their suggestion that these data were '. . . somewhat damaging to the pressure flow hypothesis' cannot be sustained.

Table 9.2 RATIOS OF VARIOUS RADIOISOTOPES SIMULTANEOUSLY APPLIED TO A LEAF AS A FUNCTION OF DISTANCE OF TRANSLOCATION. (FROM SWANSON AND WHITNEY, 1953, courtesy *American Journal of Botany*)

Distance,* mm	$^{42}K : {}^{32}P$ Translocation			Distance,* mm	$^{137}Cs: {}^{32}P$ Translocation
	1.5 h	3 h	4.5 h		4 h
70	7.9	0.73	0.49	90	1.32
90	8.3	0.53	0.51	120	1.14
110	10.6	0.38	0.43	150	0.78
130	9.7	0.52	0.48	180	0.40
150	9.4	0.58	0.31		
170	10.0	0.48	0.36		
190	7.4	0.51	0.36		
210	12.0	0.45	0.42		
230	7.6	0.51	0.35		

* Distance from radioisotope application site on primary leaf to mid-point of stem segment.

Further work on the velocity of movement of different chemical species was conducted by Biddulph and Cory (1957), who applied $^{14}CO_2$, tritiated water and ^{32}P-labelled inorganic phosphate to leaves of red kidney bean plants. The velocity of each tracer (^{32}P, 86.5 cm/h; ^3HHO, 86.5 cm/h; ^{14}C, 107 cm/h) was calculated from the distance the tracers had moved down the stem during a 15 min migration time. Although [^{14}C]assimilates moved more rapidly than either [^{32}P] or tritiated water, these experiments do not bring us any nearer to the solution of the problem; indeed, attempting to measure absolute velocities, these workers were unable to take into account the time taken for the tracers to reach the phloem from the exterior of the leaf.

On the other hand, Biddulph and Cory analysed separately the tissues exterior and interior to the cambium, showing that extensive radial movement of the tracers occurred from the phloem, the amount being proportional to concentration. Although they did not seem to appreciate the possibility of this radial loss affecting the values they obtained for the tracer velocities, they did conclude that the extent of this movement threw some doubt upon the operation of a pressure flow system. However, they also pointed out that the demonstration of independent velocities need not *per se* be completely damaging to pressure flow; unless the sieve tube protoplast were completely inert, some effects upon solute movement might be expected.

What appears to be the most unexceptionable evidence to date in favour of independent velocities is contained in the work of Fensom (1972) on excised strands of *Heracleum* phloem. Simultaneous microinjection of [14C]sucrose and 42K into a single sieve element, followed by analyses of phloem sections 2 min later, revealed the 42K to be moving at a speed of 30–60 cm/h while the 14C travelled at about 200–500 cm/h (*Figure 9.4*). Although Fensom's transport system was far less complicated than those of the whole plant investigations already described, phloem strands are still comparatively complex structures. It is not possible, therefore, to be completely certain that his results were not influenced by the activities of parenchyma cells lying adjacent to the sieve tubes.

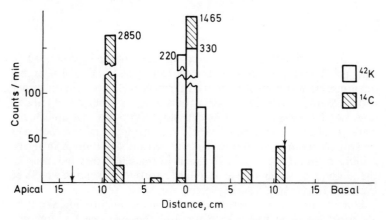

Figure 9.4 Microinjection of single sieve tubes in intact phloem strands. Histogram of counts/min per cut section when [14C]sucrose and 42KCl were simultaneously injected. Strands were frozen on dry ice 2 min after injection. (From Fensom, 1972, courtesy of the National Research Council of Canada)

It should not be expected that any startling developments will occur in this region of phloem physiology in the near future. Indeed, it is highly probable that phloem workers will tend to discontinue work on this problem, for not only is it very difficult to see how it can be successfully tackled in the contect of phloem complexity, but also even unquestioned data on different velocities are open to a wide variety of interpretations.

10

The osmotic properties of sieve elements and water movement in phloem

It is an essential feature of certain of the hypotheses described in Chapter 7 (pressure flow and electro-osmosis) that considerable quantities of water must be moved through the sieve tubes in order to transport solutes. Conversely, the surface movement hypothesis does not have such a requirement, since with this mechanism the water could be completely static. The 'protoplasmic' hypotheses present something of an intermediate picture, for though in trans-cellular streaming a mass movement is envisaged through strands, the amount of water transport relative to that of solutes would be less than in pressure flow or electro-osmosis.

Solution flow in an electrokinetic mechanism is brought about by electrical forces at the sieve plate, and the evidence for potential gradients across this structure has already been described. Pressure flow, however, is driven by gradients of hydrostatic pressure between the ends of the sieve tube. It is therefore necessary in the present chapter not only to decide whether bulk-flow mechanisms are viable on the basis of the experimental evidence for concomitant water and solute fluxes, but also to consider whether there is sufficient evidence of the presence of conditions necessary for the production and maintenance of turgor gradients in phloem.

The osmotic properties of sieve elements

Plasmolysis in mature sieve elements

There are considerable concentrations of osmotically active solutes in sieve tube saps (Chapter 2); and although mature sieve elements do not seem to possess a tonoplast, most cytologists have no doubt that a plasmalemma is present in these cells. The conditions necessary for the generation of turgor pressure would, therefore, appear to exist. However, the ease with which solutes can escape laterally from the sieve tubes (Biddulph and Cory, 1957) has cast some doubts upon the osmotic efficiency of the plasmalemma.

It was thus essential, if the pressure flow proposal was to remain viable, unequivocally to demonstrate the plasmalemma of sieve elements to be a membrane of relatively low permeability to the solutes contained in the lumen. The easiest way of doing this, clearly, was to show the sieve elements to be capable of plasmolysis. This might seem to be a relatively simple exercise, but in fact it was not until relatively recently that unassailable evidence of plasmolysis was obtained. A little reflection provides us with the reasons for this delay; not only are the sieve elements notoriously sensitive to manipulation, but to transport solutes they must have a low lateral permeability coupled with a very high longitudinal permeability. Sieve elements in sections of phloem would therefore be impossible to plasmolyse if the plasmolytic solute could readily enter via the sieve plates, as must occur with the structural conditions necessary for pressure flow, unless a solute-impermeable barrier was produced by the plugging of the sieve pores during manipulation.

First reports of sieve element plasmolysis were by Schumacher (1939), who investigated phloem from a wide variety of woody and herbaceous species, and by Rouschal (1941), working on tree phloem. Undoubtedly, however, the most detailed investigation of sieve element plasmolysis is that of Currier, Esau and Cheadle (1955), working on cotton and vine. Longitudinal sections of the phloem from these species, on average 55 μm thick, were taken from tissues which had been stored for a short time in buffered 0.25M sucrose. Concave plasmolysis forms in mature sieve elements (i.e. those lacking a nucleus) were observed when the sections were irrigated with plasmolytica composed of 0.5–2.0M solutions of glucose or fructose. A notable achievement of Currier and his colleagues was the demonstration that their procedures did not produce plasmolysis as a result of extensive damage to the sieve elements, for they were able to perform several cycles of plasmolysis–deplasmolysis on the same cell.

It is of interest to note that Currier *et al.* found it easier to plasmolyse sieve elements from dormant phloem than from active phloem. This could well have been caused by a lower permeability of the plasmalemma in dormant cells, or to more effective closure of the sieve pores in material in the dormant condition.

Osmotic potential gradients

The generation of a turgor gradient within sieve tubes in the direction of movement necessitates a higher osmotic potential of the sieve tube sap at the source than at the sink end of the system. Using this fact as a basis for experiment, Curtis and Schofield (1933) showed higher osmotic potentials in the growing (sink) tissues of onion than in the storage (source) tissues. Taking this in conjunction with the observation that the source tissues were frequently more flaccid than the sink organs, they concluded that a turgor gradient could not exist in the direction of transport.

However, since Curtis and Schofield were unable to measure the osmotic potentials of the sieve tube sap, and in view of the active nature of solute-loading processes, their inferences can hardly carry a great deal of weight. Particularly damaging to their conclusions is the observation of Currier *et al.* (1955) that strong osmotic solutions are necessary to plasmolyse sieve elements, and the experiments of Weatherley *et al.* (1959), which showed that the sieve tubes can maintain turgor in the face of considerable water stress.

Despite the fact that Tingley (1944) was unable to show the existence of osmotic gradients in stems of *Cucurbita*, the weight of evidence now supports the observations of Huber and his colleagues (1937), who demonstrated a lower concentration in exudates from the base than from the crown of trees.

Some of the most impressive data have been gathered by Zimmermann (1957b), using the incision technique to obtain sieve tube exudates from white ash. During the summer, when the trees were in full leaf and presumably assimilating and transporting, positive gradients of the four main carbohydrates present in the sap were found in a downward direction (*Figure 10.1a*).

Having established this, Zimmermann then addressed himself to the problem of what should happen to these gradients, assuming pressure flow to operate, when the supply of carbohydrates to the phloem was cut off by leaf fall in the autumn. Surely, he reasoned, flow should continue as long as a turgor difference exists between the top and the base of the tree? If there is only one osmotically active solute present, movement will cease when its concentration reaches the same value throughout the system. On the other hand, if there are several solutes having different concentration gradients, then the substances with the smallest gradients would have to be moved against these gradients until the total gradient ceased to exist (i.e., of course, if the total concentration gradient reflects the gradient of turgor pressure).

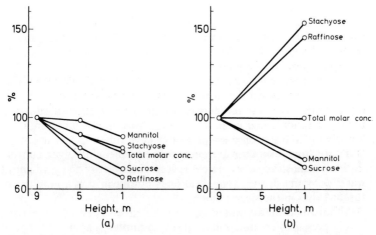

Figure 10.1 Sugar concentration gradients in sieve tube exudate of white ash. Concentrations are shown in % of the 9 m values. (a), During period of full leaf (July 30); (b), after leaf fall (October 8). (From Zimmermann, 1957b, courtesy of the American Society of Plant Physiologists)

In fact, Zimmermann found these predictions to be completely fulfilled. The sieve tubes of white ash can be tapped for a few days after leaf fall, and *Figure 10.1(b)* shows the data from an experiment in which samples were taken from a tree whose leaves had all dropped. Two of the carbohydrates, stachyose and raffinose, had a negative gradient downwards; the other two, sucrose and mannitol, a positive one; and the total molar gradient was zero. Although Zimmermann did not measure the gradients of other osmotically active solutes (potassium in particular; Peel and Weatherley, 1959), his data are cogent evidence against any hypothesis which considers independent solute movement to be possible, and are in close accord with the principles of pressure flow.

Turgor pressure gradients

Although Zimmermann's work is excellent evidence for the existence of osmotic potential gradients in phloem in the presumed direction of movement, his data do not necessarily mean that a turgor gradient also exists in the same sense. The hydrostatic pressure which can be generated by an osmotic solution when confined within a semipermeable membrane depends not only upon the osmotic potential of the solution but also upon the water potential in the external milieu. The sieve tubes lie adjacent to the xylem vessels which conduct water from roots to leaves along a gradient of hydrostatic pressure, this being highest at the base of the plant and lowest in the leaf veins. Thus, the water potential in the xylem is lowest in the upper parts of the plant.

Now, experiments with aphid stylets (Chapter 5; Weatherley *et al.*, 1959) have shown that lowering the water potential in the xylem (by osmotica) leads to an increase in the concentration of sugars in the stylet exudate and a fall in volume flow rate of exudation, i.e. a lowering of the turgor pressure in the sieve tubes. Could it not be, therefore, that in a transpiring tree the gradient of sugar concentration in the sieve tubes merely reflects the effect of the gradient of water potential in the xylem, and does not mean that a gradient of turgor exists in the sieve tubes?

Although osmotic potentials are relatively easy to measure, turgor pressures are not, and it must be admitted that at the moment we have no satisfactory answer to this question. One fact, however, is fairly certain: the sieve tube can generate considerable internal hydrostatic pressures. Mittler (1954) applied the Poiseuille formula to flow through the stylet canal of *Tuberolachnus*, after measuring the radius of the canal, the length of the canal, the flow rate and sugar concentration of the exudate. He arrived at a figure of around 15 atm—a value subsequently confirmed by Peel (1959), who gave a range of values for different stylets between 12 and 30 atm.

The study by Kaufmann and Kramer (1967) on the osmotic relations of phloem from *Liriodendron* and red maple gave no support to the concept that turgor gradients exist in the phloem of these species. Discs of phloem were removed from the trunks of trees, water potential being measured by a psychrometric technique, and osmotic potential determinations on expressed sap were carried out. Turgor pressure was calculated as the difference between water and osmotic potential. While a gradient of osmotic potential was observed in a downward direction, no turgor pressure gradient was found (*Table 10.1*).

Hammel (1968), using a direct technique for turgor pressure determinations which involved tapping the sieve elements with a stainless steel tube attached to a manometer, has produced some data from red oaks on turgor pressures and osmotic potentials. Phloem turgor pressures varied considerably in a series of measurements taken at an upper and lower level on the trunk, this variation being greater than the variation in the values of the osmotic potential. However, in a series of turgor pressure measurements the values in a sequence from the upper level were, generally, a little higher (0–3 atm) than the values from the lower level, the osmotic potential

Table 10.1 GRADIENTS IN WATER POTENTIAL, OSMOTIC POTENTIAL AND TURGOR PRESSURE IN RED MAPLE TREES SAMPLED BETWEEN 7 AND 10 p.m. (FROM KAUFMANN AND KRAMER, 1967, courtesy *American Society of Plant Physiologist*)

Sample	Leaf (6 m)	Phloem (5 m)	Phloem (1 m)	Soil
WATER POTENTIAL				
Tree A	− 4.4	− 5.4	− 3.4	−0.7
Tree A	− 4.3	− 3.7	− 2.3	−0.1
Tree B	− 5.3	− 4.4	− 2.9	−0.1
OSMOTIC POTENTIAL				
Tree A	−14.6	− 9.3	− 7.6	
Tree A	−14.5	− 8.9	− 7.5	
Tree B	−13.4	−10.0	− 8.3	
TURGOR PRESSURE				
Tree A	+10.2	+ 3.9	+ 4.2	
Tree A	+10.2	+ 5.2	+ 5.2	
Tree B	+ 8.2	+ 5.6	+ 5.4	

Values in bars.

measurements following the same pattern (*Table 10.2*). Hammel's conclusions were that translocation is favoured by a small turgor pressure gradient, but he emphasised the difficulties inherent in measuring gradients in an elastic, low-resistance distribution system composed of connected longitudinal conduits.

Table 10.2 TURGOR AND OSMOTIC PRESSURES IN SECONDARY PHLOEM OF TRUNK OF RED OAK AT AN UPPER $(6.3 \pm 0.3$ m) AND A LOWER $(1.5 \pm 0.3$ m) LEVEL. (DATA FROM HAMMEL, 1968, courtesy *American Society of Plant Physiologists*)

Date, weather, air temperature	Level	Time	Osmotic pressure	Turgor pressure
Oct. 25; overcast, after 2 p.m. rain; 17–18°C	Upper	16.15	−24.3, 24.0, 23.7	19.5, 18.0, 17.6, 16.9, 16.7, 16.1
	Lower	15.35	−23.2, 22.9, 22.0	17.4, 16.7, 14.5
Oct. 26; bright sun, breeze; 12–13°C	Upper	16.30	−24.1, 23.8, 23.1	17.2, 14.3, 12.9, 12.0, 11.4, 11.2
	Lower	15.30	−22.1, 22.1 21.5	17.6, 13.3, 12.5, 11.0, 8.2, 7.4
clear sky; 7–7.5°C	Upper	22.50	−25.1, 23.9, 23.7	19.1, 18.1, 15.6, 15.3, 15.3, 12.2
	Lower	22.00	−23.2, 22.6, 22.4	19.1, 19.0, 12.2, 11.8, 11.6, 9.5
Oct. 27; clear breeze; 13–14°C	Upper	15.10	−24.5, 24.1	18.6, 18.3, 17.9, 17.9, 14.4
	Lower	14.15	−22.9, 22.0	16.7, 16.7, 14.7, 12.7, 12.3, 9.5, 8.2

All pressures in atmospheres.

Induced turgor gradients and transport

Whatever the actual state of affairs in translocating phloem, there is another approach to the problem of the relationship of turgor gradients to movement which entails measuring the effects of induced turgor gradients on transport in isolated pieces of tissue. Rouschal (1941) observed a surging flow within files of sieve elements of *Aesculus* in the direction of a glycerine solution which bathed one end of the tissue. Bauer (1953) has studied the effect of imposed turgor gradients upon the movement of fluorescein in *Pelargonium* petioles. According to this worker, both the rate and direction of fluorescein transport could be controlled by turgor gradients.

In a more recent publication Peel and Weatherley (1963) investigated the effect of raising the hydrostatic pressure in the xylem of stem segments of willow upon exudation from stylets of *Tuberolachnus*. Initial experiments showed that a uniform increase in xylem pressure raised the turgor pressure within the pierced sieve element, as evidenced by a rise in the volume flow rate from the severed stylets, i.e. an effect opposite to that produced by introducing osmotic solutions into the wood (Weatherley *et al.*, 1959).

It was therefore argued that if a hydrostatic pressure gradient were established within the xylem, this would cause a greater flux of water into the sieve tube at the high-pressure end of the segment, and, hence, a higher turgor pressure at this end, than at the low-pressure end. Thus, on the pressure flow hypothesis, if a stylet were established at one end of a segment and the hydrostatic pressure raised at the other end, then the rate of exudation from the stylet should rise. In fact, this was shown to occur, both in intact segments and also in segments with severed xylem, in which no possibility existed of a rise in pressure in the xylem adjacent to the stylet (*Figure 10.2*).

Superficially this result seems in excellent accord with the pressure flow mechanism, since it suggests that appreciable increases in the rate of transport can be induced by modest increases in the turgor gradient (the increase in turgor pressure would be less than the increment of xylem pressure, owing to the concomitant dilution of the sieve tube sap and the fall in its osmotic potential).

However, it is most doubtful whether such a simple interpretation of the data is justified, for in other experiments described by Peel and Weatherley in which stylets were sited at the high-pressure end of a segment the rate of exudation declined with a rise of pressure. This means that there was a fall in turgor of the sieve tube in a region where the water potential had been raised. Also, it was shown in experiments in which stylets were situated at the low-pressure end of a segment that the longer the segment, the greater was the increase in exudation upon application of pressure, an effect difficult to reconcile with a simple effect on a pressure flow mechanism, where it would be expected that increases in the rate of stylet exudation should be proportional to the steepness of the extra pressure gradient, i.e. the shorter the segment, the greater the rise in exudation rate.

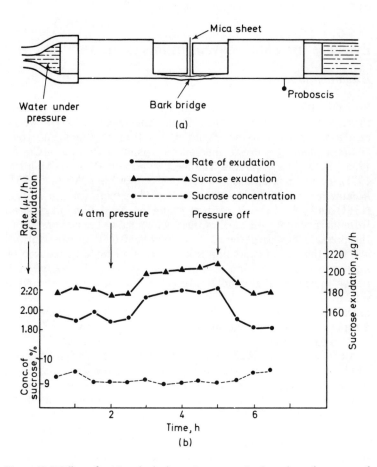

Figure 10.2 Effect of raising the hydrostatic pressure in the xylem of one part of a segment on exudation from a stylet situated on another part, in the xylem of which the pressure was not raised (b). The two parts of the segment were connected only by a bark bridge (a). (From Peel and Weatherley, 1963, courtesy of The Clarendon Press)

In fact, Peel and Weatherley concluded that the effects of pressure gradients in the xylem were somehow bound up with changes in the length of bark which contributed solutes to the stylets, an inference subsequently substantiated by the work of Peel (1965b). While it is possible to explain this effect in a very general way, it is clear that pressure gradients do not act altogether by simply enhancing a pressure flow system. Some of their effects are indirect and complex; thus it is not possible to conclude from these experiments that solute transport in willow is mediated via pressure flow.

Evidence for water movement in sieve tubes

Exudation processes

Probably the most cogent evidence for bulk flow of a solution through the sieve tubes comes from those experiments in which exudation of a solute-rich liquid occurs from incisions into trees (Zimmermann, 1957a) or herbaceous plants (*Cucurbita*—Crafts, 1936; *Yucca*—van Die and Tammes, 1966; and *Ricinus*—Milburn, 1971), or from severed aphid stylets (Kennedy and Mittler, 1953; Weatherley *et al.*, 1959). Indeed, it was the experiments of Hartig, (1861) and of Münch (1930), working on exudation phenomena from incisions into the bark of trees, which led to the elaboration of the pressure flow hypothesis. Most pertinent of their observations was that the rate of exudation from an incision could be reduced by making a second incision several metres above or below the first, i.e. an indication that exudation involves the movement of a solution towards the site of injury. Although this effect could not be repeated by Moose (1938), Zimmermann (1960b) has demonstrated a rapid effect on the concentration of the exudate from a distance as far away as 10 m.

Although there is no doubt that exudation from incisions involves some water movements (i.e. at least through the incised elements), we cannot as yet answer the following questions with any degree of confidence: (a) Is the water which issues from a cut moving rapidly through the sieve plates? (b) If it is, then does a similar water flux take place in the undisturbed sieve tubes? While it appears highly probable that solution flow occurs through sieve plates after an incision has been made, it must not be forgotten that opening to the atmosphere an elastic system containing fluid under a high positive pressure will inevitably cause a sudden rush of the contents towards the wound. It is not difficult to envisage the situation in which normally static structures (i.e. plasmatic filaments) are forced out of the sieve tubes by the large pressure gradients generated; indeed, Zimmermann (1957b) describes exudation as an 'explosion-like' process, an expression not likely to induce confidence in the relevance of the phenomenon to processes occurring in undisturbed phloem.

Incision-exudation is usually a short-lived process (e.g. in *Ricinus*, *Figure 10.3*), although it can continue for as long as 24 h in *Yucca* (van Die and Tammes, 1966). The most likely explanation for the transient nature of exudation from incisions is that the sieve elements at the surface of the cut become plugged by one or more of the processes described in Chapter 6, evidence for this being gained from the fact that exudation can frequently be restarted by the removal of a thin slice of tissue from the surface of the wound. There is also no doubt that incising sieve elements causes a rapid decrease in their turgor, a consequent influx of water due to the change in diffusion potential gradient, and thus a rapid dilution of the contents of these cells. Zimmermann (1957b) has provided some very clear data (*Figure 10.4*) showing this dilution effect. Collapse of the injured sieve elements, due to an inability of surrounding cells to pump solutes into them at a high enough rate, may therefore be another reason why incision-exudation does not proceed for extended periods.

What seems to be the best current method of studying sieve tube transport by tapping techniques is the use of severed aphid stylets; although some damage is done to the extracambial tissues during stylet insertion, little if any harm is done to the sieve elements, as evidenced by the fact that exudation can proceed for periods of up to a week (Peel, 1959). Of course, the food canal of the stylet bundle is of very small diameter (1–2 µm in *Tuberolachnus salignus*); therefore, the drop in sieve element turgor and the displacement of its contents will be much less severe than with large incisions. Sealing mechanisms will not thus be triggered into operation by stylet puncture as quickly as by large incisions.

In a paper published in 1959 Weatherley and his colleagues described a number of experiments, many of which were designed to show that exudation from stylets bore some relevance to normal sieve tube transport. Although it seemed likely from a consideration of permeability data that exudation involved a degree of longitudinal transport, the possibility that both water and solutes were entering the pierced sieve element by diffusion over its whole surface could not be completely excluded.

Repeating the incision experiments of Münch (1930), Weatherley *et al.* failed to find any marked effect upon exudation of a girdle interposed on a willow cutting between a stylet and the leaves, from which they inferred that there must be a switchover in the source of solutes, from the leaves to the storage cells of the stem, when the former supply is cut off by a girdle. In fact, the stylets will exude when sited upon isolated segments of stem when these are supplied with water, and it was while using isolated segments that Weatherley *et al.* were able to confirm the supposition that stylet exudation involved longitudinal movement across sieve plates.

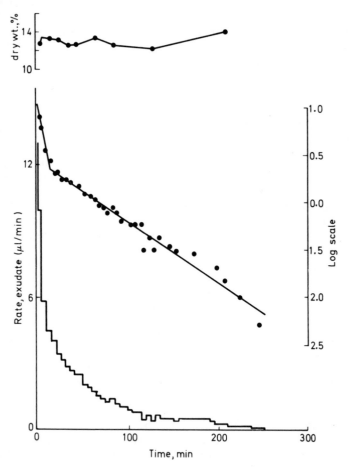

Figure 10.3 Exudation profile from a vigorously growing plant of Ricinus *(lower curve); logarithmic plot of same profile (middle curve); and the % dry wt. of the exudate collected throughout (upper curve). (From Milburn, 1971, courtesy of Springer-Verlag)*

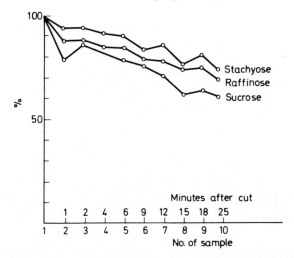

Figure 10.4 Osmotic dilution effect in sieve tube system of white ash. Subsequent samples taken from one tapping cut show a continuous decrease in concentration. The cut yielded sap for about half an hour. Sugar concentrations shown in % of the concentrations in the first sample, which were: stachyose 0.242M, raffinose 0.077M, sucrose 0.068M. (From Zimmermann, 1957b, courtesy of the American Society of Plant Physiologists)

After siting a stylet near to the basal end of a segment 40 cm in length, they measured the effect upon exudation of transverse razor cuts made into the bark at positions progressively nearer to the stylets. When incisions were made at a distance of less than 16 cm from the stylets, there was found to be an immediate slowing in the rate of exudation, followed by a partial recovery to a lower steady rate. The results of a number of experiments showed clearly that some of the exudate must have reached the stylets from a distance of 16 cm, i.e. it must have traversed between 800–1000 sieve plates. It was this distance of 16 cm which Peel and Weatherley (1962) termed the 'contributory length'.

In this latter paper Peel and Weatherley attempted to provide an analysis of the stylet exudation process. After the stylets have pierced a sieve element, they suggested, leakage through the food canal leads to a fall in turgor pressure in the punctured element, and this in turn induces two fluxes: one an influx of water and solutes through the walls of the sieve element from surrounding cells (f_1 in *Figure 10.5*), the other a flux (F_1) of solution through the sieve plates from the contiguous elements (S_2). Thus, elements S_2 will in turn suffer a leak (F_1) which causes a twofold flux (f_2 and F_2) into them. In this way stylet puncture causes a disturbance to spread along the sieve tube in both directions from the point of puncture, and under steady conditions f_1 and F_1 will have the highest values and the series f_2, f_3, etc., and F_2, F_3, etc., will die away on either side until a point is reached where the cells suffer no disturbance (the latter may never be reached in practice, for Ford and Peel, 1966, have shown the contributory length to be greater than 16 cm, and may indeed be equal to the whole length of a segment).

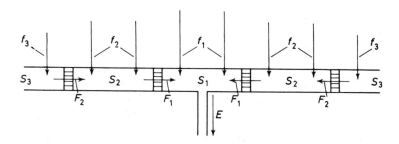

Figure 10.5 Illustration of the pattern of fluxes during sieve tube exudation from severed stylets; E denotes the flux of sieve tube sap through the food canal of the stylet bundle. (From Peel and Weatherley, 1962, courtesy of The Clarendon Press)

Peel and Weatherley considered stylet exudation only to involve longitudinal movement in one sieve tube. However, there seems to be no *a priori* reason why stylet puncture should not initiate longitudinal fluxes in sieve tubes adjacent to the one containing the pierced element, i.e. there may be a lateral as well as a longitudinal 'contributory length'.

The experimental data provided by Weatherley *et al.* (1959), combined with the plausible analysis of the process by Peel and Weatherley (1962), certainly form cogent evidence in favour of assuming that stylet exudation involves a rapid flow of solution through sieve plates down a gradient of hydrostatic pressure. Although Peel and Weatherley (1962) did not speculate upon the nature of the longitudinal fluxes (*F, Figure 10.5*), the incision experiments of Weatherley *et al.* (1959) can only be interpreted as showing that, at least, the solutes of the exudate must be moving along the sieve tube to the point of stylet puncture. Thus, if exudation does not involve bulk, longitudinal fluxes of solution, the only other possible explanation of the process must be that solutes only are pumped laterally and longitudinally into the pierced element, water following by an osmotic process (i.e. a situation analogous to ion and water movement across detopped root systems).

In addition to the evidence from exudation processes, there are other studies which support the concept that both solutes and water move together through sieve tubes. The work of Ziegler and Vieweg (1961), using a thermoelectric technique on phloem strands of *Heracleum*, has already been referred to in Chapter 4, and the interesting experiments of Münch (1930) are worthy of mention in the context of longitudinal water movement.

With a pressure flow mechanism (or an electro-osmotic one, for that matter) nutrient movement is effected by moving large quantities of a relatively dilute solution, i.e. large quantities of water must be transported per given mass of solute. Now, in an intact plant system the excess water arriving at the sink end of the system can readily be removed via the xylem. On the other hand, if transport could be induced to occur into a 'cul de sac' of phloem which is separated from the xylem, then the excess water could only escape by being exuded from the phloem tissue.

Münch levered a flap of bark away from the wood at the base of a tree, in such a way that it was in contact with the rest of the bark only at its top. Assimilates from the leaves should, therefore, move into the flap, carrying with them solvent water, if transport were a pressure flow. Thus, it should be possible to collect this water, provided that evaporation were prevented by enveloping the bark in a suitable material; and, indeed, Münch found that water dripped from the end of the bark flap.

Before we leave the topic of phloem exudation, a recent study by Tammes and Ie (1971) is of interest in the context of the size of the channels through sieve plates. Introducing this topic, Tammes and Ie quote work on the passage of viruses and newly discovered plant pathogens called mycoplasma (Doi *et al.*, 1967) through sieve plates (Giannotti, Devauchelle and Vago, 1970) in support of the thesis that open channels exist through the pores, for the smallest particles of the mycoplasma are 500 Å in diameter. By placing 3-mm-long segments of *Yucca* inflorescence stalk upon an aqueous suspension of carbon black particles, followed by electron microscopy of sections of the stalk, Tammes and Ie showed that particles of between 200 and 700 Å in diameter had been sucked into the sieve elements and moved through the sieve plates.

Studies with labelled water

Water labelled with the stable isotopes of hydrogen (^2H) or oxygen (^{18}O) is readily available, and since the advent of liquid scintillation spectrometry the radioactivity of tritiated water (^3HHO) can easily be assayed. It might appear reasonable, then, to feel that isotopic tools should be able to settle the problem of whether large quantities of water are moved through sieve tubes.

As with most techniques, however, the data produced by experiments with labelled water present many difficulties of interpretation; problems arise in getting the label into the phloem and once there of confining it to that tissue, for water can move with great ease throughout the plant. Tritiated water is particularly difficult to use, since the radioactive hydrogen atoms can exchange with unlabelled hydrogen atoms on other molecules. Considering the ubiquitous occurrence of hydrogen in biological molecules, it is, therefore, easy to envisage considerable losses of radioactivity from tritiated water between the source and sink.

The use of tritiated water in studies on phloem was first reported by Biddulph and Cory (1957). Some of the results of their work have been discussed in the preceding chapter, but in the context of water transport the main inference to be drawn is that water is longitudinally mobile in the phloem. Tritiated water was supplied to the under-surface of a terminal leaflet of a trifoliate leaf of bean; subsequent analyses of stem sections below the treated leaf after a maximum migration period of 30 min showed tritium activity to have moved a considerable distance from the application site.

Since Biddulph and Cory also investigated the simultaneous transport of ^{14}C-labelled sugars, they were able to obtain an estimate of the ratios of labelled sucrose and labelled water exported from the leaves. In fact, they showed that the extent of tritiated water transport was only one-fortieth of that required to move the sucrose as a 10% solution. However, they did not feel that this was evidence against a bulk flow of a relatively dilute sugar solution through the phloem, for, as they point out, the low value for labelled water movement might have been due to a lack of equilibration between the labelled and unlabelled water in the leaf.

Quite opposite conclusions were reached by Gage and Aronoff (1960) and by Choi and Aronoff (1966), working on soyabean. In the latter study tritiated water vapour was fed to a single leaf held in a water-saturated atmosphere in the light (all other leaves having been removed), after which the tritiated water profile was shown to extend but a short distance down the petiole from the leaf. When darkened petioles were exposed to tritiated water vapour, the rest of the shoot being kept in the light, almost half the activity found in the tissues had moved acropetally. Only 10% moved basipetally in a migration period of 1 h.

Choi and Aronoff attempted to use their data to determine whether transport of sucrose took place by bulk solution flow, by comparison with two model systems: one in which self-diffusion of tritiated water was considered with water and sugars moving independently (tritium photosynthetically fixed in sugars was assayed along with tritiated water), and a bulk flow model in which water exchanged freely between the phloem and surrounding tissues. They concluded from this analysis that bulk flow was not a dominant process in soyabean translocation.

Further investigations by Trip and Gorham (1968b) on *Cucurbita* indicate that the relative mobilities of tritiated water and ^{14}C-assimilates can be influenced by the tracer application procedure. No evidence of concurrent translocation was seen when ^{14}C was administered as $^{14}CO_2$ and tritium by painting the surface of a leaf with tritiated water. On the other hand, when ^{14}C-labelled glucose and tritiated water were applied to the leaf via a flap formed by severing a lateral vein, concurrent translocation of the two tracers took place, as evidenced by the constancy of their ratios with increasing distance down the petiole (*Table 10.3*).

Table 10.3 DISTRIBUTION OF ^3H AND ^{14}C IN 0.5 cm SEGMENTS AFTER VARIOUS PERIODS OF TIME FOLLOWING FLAP-FEEDING OF [^{14}C]GLUCOSE AND ^3HHO. (FROM TRIP AND GORHAM, 1968b, courtesy *American Society of Plant Physiologists*)

Segment	1 h $^{14}C : ^3H$	2 h $^{14}C : ^3H$	3 h $^{14}C : ^3H$
Fed side vein	0.14	0.05	0.18
Supply leaf blade	0.15	0.06	0.08
Upper petiole		2.2	1.3
Upper middle	5.4	1.8	1.3
Lower middle	6.5	1.6	1.3
Lower petiole	5.8	2.3	1.0

Although Trip and Gorham failed to locate the tissues responsible for labelled water movement, using autoradiography, they showed that steam girdling the petiole blocked the movement of both ^{14}C and ^3H. They thus concluded that water moved through the phloem with the sugars, which indicated transport to be mediated via solution flow.

The most recent reports of studies using tritiated water have been by Cataldo, Christy and Coulson (1972), Cataldo, Christy, Coulson and Ferrier (1972) and Fensom (1972). In the first-mentioned of their papers Cataldo and his co-workers devised a mathematical model in which reversible exchange of tritiated water between the sieve tube and surrounding cells was used to explain the difference in the apparent velocities between labelled water and assimilates (Biddulph and Cory, 1957). In the second paper they studied translocation profiles of tritiated water and [^{14}C]sucrose in the petiole of sugar beet after application of these tracers through a lateral leaf vein by a 'flap' technique. Although different apparent velocities were found (tritiated water moving more slowly than [^{14}C]sugars), Cataldo *et al.* considered, in view of their mathematical analysis, that this was brought about by different degrees of lateral exchange, an inference supported by the effect of localised low temperatures upon the exchange characteristics of the two tracers.

Whether the arguments of Cataldo and his colleagues are correct or not, the undisputed fact remains that they were able to obtain a considerable degree of labelled water movement. Similarly, Fensom (1972) has demonstrated movement of tritiated water in displaced phloem strands of *Heracleum*, when the tracer (together with others) was applied to a restricted portion of the surface of the strand.

Some work by Anderson and Long (1968) also deserves mention in the context of water movement in phloem. These workers measured the rates of transfer of tritiated water along excised roots of maize. They showed that movement took place both acropetally and basipetally with equal velocity, from which they inferred that it occurred in the phloem. Moreover, the rate of movement in normal roots was some four times greater than in roots killed by boiling, or poisoned with metabolic inhibitors. This observation, they argued, was indicative of a long-distance bulk flow in the sieve tubes, possibly brought about by movement in transcellular strands.

All the work with labelled water described so far has been conducted with intact phloem systems. Although there is much to commend this approach, it might seem rather surprising that investigators should not have used exudation processes to study water movement; this process involves the issue of water out of the phloem, and surely here it should be easy to demonstrate longitudinal fluxes of labelled water. However, the only reports on these lines seem to be by Peel *et al.* (1969) and Peel (1970).

In the first of these publications a double-chambered bark strip system was employed, an exuding aphid stylet being established on the bark under each chamber. ^{32}P-labelled phosphate, ^{14}C-labelled sucrose and tritiated water were introduced into one chamber, and the movement of these tracers into the exudate from both stylets was monitored. All three moved rapidly into the exudate from the stylet immediately below the site of application, tritiated water reaching a high level within 1 h. However, although ^{32}P- and ^{14}C-activities were detected in the exudate from the other stylet within 1 h from application, tritium activity did not appear until at least 4 h had elapsed, even though this stylet was only 1–2 cm distant from the labelled solution.

The conclusion of Peel and his colleagues that this was good evidence against a mass flow of solution has been strongly criticised by Crafts and Crisp (1971), mainly on the basis that tritiated water movement was slowed down owing to extensive lateral exchange. This could indeed be so, but such exchange must have been severe, for Peel *et al.* also showed that tritiated water could move appreciable distances by diffusion in dead pieces of bark, at a speed equal to that in live bark on which stylets were exuding. It also seems strange that so much exchange of tritiated water took place in the bark tissues, which would have a high water potential, whereas other workers have obtained rapid longitudinal transport of labelled water in transpiring plants; in the latter extensive radial movement of activity from the phloem to the xylem might have been expected, in view of the low water potentials prevailing in the xylem tissues.

In the second report Peel (1970) attempted to investigate the problem of water movement in a manner which precluded lateral loss of tritium activity affecting the results. Gradients of tritiated water, [^{35}S]sulphate and [^{32}P]phosphate were established in an isolated stem segment, a single aphid being then sited at the high-activity end (*Figure 10.6a*). Thus, the aphid would excrete honeydew containing all three tracers, but the activities of those in the honeydew should be 'diluted' by unlabelled solutes and water from the low-activity end of the segment if a bulk flow occurred towards the stylet. Therefore, if the supply of unlabelled water or solutes to the stylet is removed by girdling the segment near to the aphid (*Figure 10.6a*), the specific activities of the three tracers in the honeydew should rise. On the other hand, if the aphid were not receiving unlabelled water or solutes from a distance, then the specific activity of the labelled compound in the honeydew should remain constant.

In fact, Peel showed that while the specific activities of both ^{35}S and ^{32}P in the honeydew rose after girdling, tritium did not (*Figure 10.6b*). Thus, it would appear that puncture of a sieve element of willow by a stylet does not induce a rapid longitudinal flux of unlabelled water along the sieve tube.

It has been shown that sieve elements possess the necessary apparatus for the development of turgor pressures, and high pressures patently do exist in these cells. Gradients of osmotic potential are present in translocating phloem in the direction of movement, but concomitant gradients of turgor pressure have not yet been demonstrated. The weight of evidence, from both experiments and theoretical considerations, is in favour of transport being mediated via a bulk flow of solution, although completely unequivocal evidence for rapid water movement through sieve plates is not yet available, and certain lines of research require to be continued further before a final conclusion can be made. Thus, the results of work described in this chapter support pressure flow, electro-osmosis and certain 'contractile' mechanisms, with the latter two proposals being most strongly favoured.

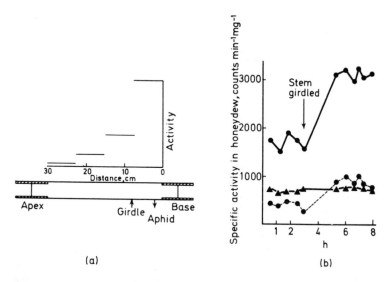

Figure 10.6 (a) Diagram illustrating the procedure used to study water movement along the sieve tube towards the site of puncture by an aphid stylet. (b) Effect of girdling a segment on the specific activities of tritium (triangles, solid lines), 35S (closed circles, dashed lines) and 32P (closed circles, solid lines) in the honeydew. (From Peel, 1970, courtesy of the Scandinavian Society of Plant Physiology)

11

Naturally occurring growth regulators, their movement and effects upon nutrient transport

Since the pioneering work of Went in 1928, a very considerable amount of attention has been directed towards an elucidation of the mechanisms responsible for the transport of naturally occurring growth regulators, particularly indole-3-acetic acid (IAA). The literature on hormone movement, therefore, is extensive, and it is not possible here to provide more than a brief outline of the subject. However, there are several recent reviews (McCready, 1966; Goldsmith, 1968, 1969) to which the interested reader is referred.

A great deal of work on the transport of hormones has, quite understandably, been performed with a view to clarifying the role of these substances in the control of various growth and differentiation phenomena. Furthermore, since many of the studies have been carried out on short segments of tissue which frequently lacked well-differentiated vascular strands, studies in the growth regulator transport field have tended to develop on lines separate from those concerned with long-distance movement of nutrients. However, recent investigations have demonstrated not only that the indole auxins, gibberellins and cytokinins can move in vascular tissues, but also that some of these substances markedly influence the distribution patterns of nutrients such as sugar. Such observations will undoubtedly lead to a synthesis of approach and thought in the two fields of hormone and nutrient transport.

The transport of growth regulators

Indole acetic acid (*IAA*)

The classical method (Went, 1928; van der Weij, 1932) for studying the transport parameters of auxin consists of applying the growth substance in an agar block (the donor) to a short segment of tissue, and collecting it at the other end in a second agar block (the receiver). Prior to the advent of radiotracers the quantity of auxin moving into the receiver blocks had to be measured by bioassay techniques, but in recent years extensive use has been made of IAA labelled with ^{14}C in either the carboxyl group or the ring. It is then a relatively simple procedure to assay the radioactivity present in the reciever block after a known period of transport.

Radiotracer methods are, of course, based upon the assumption that the activity present in the receiver block is in IAA and has been transported as IAA. The available evidence certainly justifies such an assumption. Quantitative extraction of the label from receiver blocks, followed by chromatography in several different solvent systems with a marker of known IAA has demonstrated that more than 90% of the activity in these blocks co-chromatographs with the IAA sample (Naqui and Gordon, 1965; Gorter and Veen, 1966). Furthermore, Pilet (1965), using sections of *Lens culinaris* stems, has obtained results with carboxyl- and ring-labelled IAA which were quantitatively identical, indicating that this tissue does not decarboxylate IAA to any marked extent.

Probably the most outstanding feature of IAA transport in isolated sections of tissue is that it exhibits marked polarity: more moves basipetally than acropetally, Although there is a distinct possibility that some movement of auxin in isolated tissue sections takes place by diffusion, there is a considerable body of evidence which supports the idea that there is a thermodynamically active component in the basipetal transport process (Goldsmith, 1969).

Certainly, there is little doubt that basipetal IAA transport is dependent upon metabolism: movement is inhibited by a variety of metabolic inhibitors such as cyanide, 2,4-dinitrophenol, iodoacetate and azide (Niedergang-Kamien and Leopold, 1957), and by reduced oxygen tension (Goldsmith, 1966). A number of substituted phenoxy acetic acids such as 2,4-D, as well as substances such as tri-iodobenzoic acid (TIBA) and ethylene (Goldsmith, 1969), also reduce transport.

IAA can move basipetally against a concentration gradient. Van der Weij (1934) demonstrated that IAA continued to move into receiver blocks even when its concentration in them was greater than in the donor blocks. However, as pointed out by Goldsmith (1969), this may not be indicative of an active transport process, for it is not known whether IAA is transported in a charged, ionised form or not. If movement occurs in the ionised condition, then it would have to be shown that an electrical potential gradient did not exist if the strict requirements of an active process were to be met (Dainty, 1962).

Some controversy exists as to whether the mechanisms of acropetal and basipetal transport are fundamentally different; polarity could, clearly, be expressed by a small, non-metabolic component in both directions, basipetal movement being enhanced in relation to acropetal by the intervention of a large metabolic component. The disappearance of polarity in inhibited oat coleoptiles (Goldssmith, 1966) could well be taken as evidence for the view that here the basipetal metabolic component ceased to function, leaving the non-metabolic processes in both directions still working. On the other hand, De la Fuente and Leopold (1966) have put forward a different proposal. They suggest that their data showing a logarithmic increase in polarity with length of section are indicative of a metabolic component in both directions.

Polarity of IAA movement occurs not only in sections taken from tissues such as coleoptiles and pith which lack vascular strands, but also in the vascular tissues themselves (veins of tobacco leaves—Avery, 1935). However, it seems most unlikely that movement in the short (5–20 mm) sections used in most work on IAA involves the sieve elements, since IAA transport in these sections displays several differences from nutrient movement in phloem. It has already been seen in Chapter 5 that sieve tube transport does not exhibit any detectable polarity in the transport of nutrients. Moreover, the measured values of the velocity of IAA and other growth regulator movement in short tissue sections is much lower (*Table 11.1*) than the velocities generally found for assimilate movement in sieve tubes of intact plants, being much closer to those found for radial and tangential sugar movement (Chapter 1).

It seems likely, then, that growth regulator movement in tissue sections involves a cell-to-cell mechanism, for which Goldsmith (1969) visualises two possible routes. In the first, auxin would pass through the plasmalemma of the first cell into the symplast, thence across the cytoplasm and out through the plasmalemma into the apoplast before entering the second cell. The second route would be entirely symplastic except for the initial entry from the donor and exit into the receiver blocks, movement between cells being mediated via the plasmodesmata and accelerated through the cells by cyclosis. At the present time it appears that nothing definite is known about the role of plasmodesmata in IAA transport.

Table 11.1 SOME TYPICAL VELOCITIES OF GROWTH REGULATOR MOVEMENT IN A BASIPETAL DIRECTION. (FROM GOLDSMITH, 1969, courtesy *McGraw Hill*)

Tissue	Substance	Approx. velocity, mm/h	Reference
Zea coleoptile	IAA	14	Hertel and Leopold (1963)
Zea root	IAA	6	Hertel and Leopold (1963)
Avena coleoptile	IAA	8–15	van der Weij (1932)
			Went and White (1939)
	naphthalene acetic acid (NAA)	3.7	Went and White (1939)
	anthracene acetic acid	5.3	Went and White (1939)
Helianthus epicotyl	IAA	7.5	Leopold and Lam (1961)
	NAA	6.7	Leopold and Lam (1961)
Phaseolus epicotyl	IAA	5.7	McCready and Jacobs (1963)
	2,4-D	0.6–1.0	McCready (1963)
Arachis gynophore (an intercalary meristem)	IAA	10	Jacobs (1961)

It is highly probable that IAA can move in intact plants in an analogous manner to that found in tissue sections. Morris, Briant and Thomson (1969) have reported experiments in which ^{14}C-labelled IAA was applied to stem apices of intact pea plants. Although some of the applied IAA was converted to immobile IAA-aspartate or complexed to protein, some of the activity moved throughout the plant in the form of IAA (*Figure 11.1*). However, the calculated velocity of movement was low (11 mm/h), a value hardly indicative of sieve tube transport.

On the other hand, there are a considerable number of reports which show that IAA can move into, and through, phloem. Little and Blackman (1963) studied the export of IAA applied to primary leaves of *Phaseolus*, finding velocities of movement much greater than those shown in coleoptile systems. Patterns of distribution of labelled auxin applied to leaves have been examined by Fletcher and Zalik (1965) and Whitehouse and Zalik (1967). These showed acropetal as well as basipetal movement, and in fact were virtually the same as could be predicted if the applied IAA were moving with sugars in the translocation stream.

Eschrich (1968) demonstrated that labelled IAA moved into the sieve tubes, when applied to the leaves of *Vicia*, by analysing the distribution of radioactivity in honeydew obtained from feeding aphids. Field and Peel (1971b, 1972) and Lepp and Peel (1971b) have shown labelled IAA to be capable of movement into, and

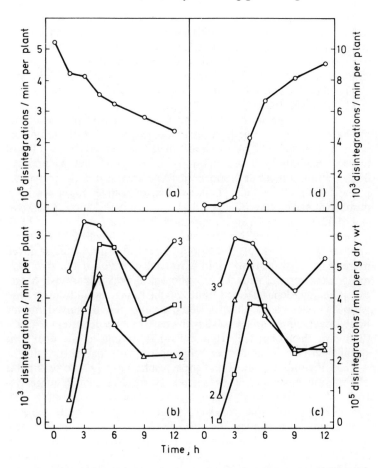

Figure 11.1 Time course of changes in the total ethanol-soluble radioactivity from the apex (a), stem (b) and roots (d) of dwarf pea seedlings labelled at the apex with IAA-2[14C]. Specific activity based on the dry weight of a parallel sample of 10 internodes is shown in (c). Numbers refer to internodes 1 to 3. (From Morris, Briant and Thomson, 1969, courtesy of Springer-Verlag)

through, the sieve elements of willow, by collecting sap from severed stylets or whole aphids sited upon isolated stem segments or bark strips. Hoad, Hillman and Wareing (1971) have not only shown that exogenous, labelled IAA can move into sieve elements but, more interestingly, demonstrated the presence of endogenous auxin activity in sieve tube exudate of willow collected as aphid honeydew. Even in short (15 mm) tissue sections of *Coleus* internodes, the label from [^{14}C]IAA becomes heavily concentrated within the phloem within 1 h from application (Wangermann, 1970). This, of course, does not necessarily mean that transport takes place through the sieve elements of these sections. Other autoradiographic studies employing tritiated IAA (Sabnis, Hirshberg and Jacobs, 1969), and ^{14}C-labelled IAA (Bonnemain, 1970), have shown localisation of the label in phloem tissue.

What would be most interesting from the point of view of the mechanism of sieve tube transport would be a demonstration of the polar movement of IAA in sieve tubes, since it would be difficult to reconcile this with certain of the hypotheses, such as pressure flow. Certainly there are reports of IAA moving in a polar manner in bark tissues of woody plants, which, of course, contain a large number of sieve tubes (Cooper, 1936, working on lemon; Oserkowsky, 1942, using apple twigs; Lepp and Peel, 1971a, employing willow).

Lepp and Peel (1971a) demonstrated that labelled IAA, when applied to the bark of 10-cm-long pieces of willow stem, moved more in a basipetal than in an acropetal direction if the stem segments were vertical with the morphological apex uppermost. Moreover, having demonstrated that labelled IAA would move into willow sieve elements, they showed that more activity moved into the honeydew from an aphid sited at the base of a vertical stem segment than into the honeydew from an aphid feeding at the apex (the specific activity of the former being up to twice that of the latter).

Thus, it could be argued that these results show IAA to be capable of polar movements in sieve elements. However, although exogenous IAA will certainly move through sieve tubes, it could well be that the data of Lepp and Peel were produced by two quite separate mechanisms: a non-polar transport in the sieve tubes upwards and downwards from the IAA application site towards the feeding aphids, the basipetal movement being enhanced by a polar system located outside the sieve tubes. Thus, more activity would appear in the honeydew from the basal aphid owing to lateral movement of labelled IAA into the pierced sieve tubes from surrounding cells.

Other growth regulators: gibberellins, cytokinins and abscisic acid

Clor (1967), using ^{3}H-labelled gibberellic acid applied to segments of pea stems or cylinders of potato tube tissue, could find no evidence of polar transport, and there seem to be no reports which present results to the contrary.

In whole plants gibberellins appear to be quite mobile. When applied to the leaves of pea seedlings, ^{14}C-labelled gibberellic acid seemed to be phloem-mobile, for it moved into immature leaves but not into those which were mature (McComb, 1964). Evidence from studies on the presence of endogenous gibberellins in phloem exudates (Kluge *et al.*, 1964; Hoad and Bowen, 1968) seems to be unequivocal evidence for the phloem-mobility of these substances, since it is most unlikely that, having entered the sieve tube, they would be longitudinally immobile. Endogenous gibberellins appear to be synthesised in the shoot tips of sunflower plants, from which site they move basipetally, or can show acropetal transport after being synthesised in leaves (Jones and Phillips, 1966). Very possibly these movements are mediated via the phloem.

Gibberellins are synthesised in the roots as well as the aerial portions. Jones and Phillips (1966) have demonstrated that this occurs in the apical 3 mm of sunflower roots. Since gibberellins have been detected in xylem saps (Phillips and Jones, 1964; Carr *et al.*, 1964), it is very possible that upward movement occurs in the xylem, although radial movement of ^{14}C-labelled gibberellic acid from the xylem to the phloem can readily occur in willow stems (Bowen and Wareing, 1969).

Cytokinins, like gibberellins, do not, generally, appear to move in a polar manner, although Black and Osborne (1965) have reported polar transport of both IAA and the kinin benzyladenine in sections of bean petioles, the extent of polarity of benzyladenine being much less than that of IAA. Wareing (1970) has described experiments performed in his laboratory, which suggest that gravity may affect the distribution of IAA and kinetin in horizontal seedlings of birch; IAA moving to the lower side, kinetin to the upper. Gravity also seems to affect labelled kinetin distribution in decapitated bean plants.

Cytokinins seem to be somewhat less mobile in whole plants than do gibberellins, at least as far as the phloem is concerned. Early experiments by Sachs and Thimann (1964) showed that while kinetin would break the apical inhibition of lateral buds of pea seedlings when applied directly to the buds, it would not do so when applied but a short distance away. Some observations by Lepp and Peel (1971a) appear to indicate that kinetin is not as mobile as IAA in the phloem of willow. However, kinetin can move in phloem, and radially between the xylem and phloem tissues (Bowen and Wareing, 1969). Endogenous cytokinins may also be present in phloem sap (Hillman, 1968, quoted by Bowen and Wareing, 1969).

Cytokinins are probably synthesised in the roots and exported via the xylem to the aerial portions of the plant. Certainly cytokinins have been detected in xylem saps from sunflower (Kende, 1965) and from *Acer saccharum* and vine (Nitsch and Nitsch, 1965), which shows that they are xylem-mobile.

The growth retardant abscisic acid seems to be mobile in both phloem and xylem. Hoad (1967) has detected this substance in phloem exudate obtained as aphid honeydew from willow cuttings grown under short-day conditions, and has suggested that it is transported from the mature leaves to other parts of the plant, where it causes the onset of dormancy. Bowen and Hoad (1968) have detected abscisic acid in the xylem sap from willow together with a second growth inhibitor. Lenton *et al.* (1968) have also shown abscisic acid to be present in xylem sap. Abscisic acid may therefore be synthesised in the roots as well as the mature leaves, and be transported via the xylem.

'Hormone-directed' transport

Plant hormones are so widespread in their effects that it is not surprising that a considerable body of evidence has been accumulated over the years which demonstrates that they can, under certain circumstances, affect the distribution patterns of various nutrients. The most usual form of experiment designed to show hormonal influence upon movement consists of hormone application to a particular organ, (1) followed after a period of time by analyses of endogenous nutrients in the organ, or (2) following the accumulation of a labelled nutrient (e.g. [^{14}Ca]assimilate or [^{32}P]phosphates) in the organ after their application at another site. By means of such procedures it has been unequivocally demonstrated that growth substances such as IAA and the kinins can affect both the direction and the rate of nutrient movement.

What is not understood, however, is how these effects are mediated. One possibility is that the applied hormones might stimulate growth of the tissues, thereby forming an active 'sink' which would readily draw upon nutrients. On the other hand, the applied hormones could move into the vascular tissues and affect transport by directly influencing the transport processes within the sieve elements. Although it is rather difficult to conceive of the hormones affecting a passive process such as pressure flow, a direct effect upon an active mechanism is not out of the question. Thimann and Sweeney (1937) have demonstrated a stimulation of protoplasmic streaming by IAA at certain concentrations, possibly mediated via an effect upon protoplasmic viscosity (Turner, Macrae and Grant-Lipp, 1954). Also, Ilan (1962) has shown IAA at a concentration of $10^{-4}M$ to stimulate potassium uptake by sunflower hypocotyl cells. Thus, IAA could possibly affect either a streaming or an electro-osmotic mechanism.

Went (1936) suggested that the phenomenon of apical dominance and the correlative inhibition of lateral buds might involve an auxin-mediated diversion of nutrients towards the growing apical regions. This view seems to be supported by the work of Mitchell and Martin (1937), who showed that accumulation of carbohydrates and nitro-genous compounds occurred in the apical regions of decapitated bean plants if these were treated with auxin, and by work in recent years which has employed radioactive tracers in short-term experiments.

In 1962 Booth *et al.* described work they had performed on the movement of ^{14}C-labelled sucrose in pea plants bearing 4–5 expanded leaves. Two series of the plants were decapitated, the cut stumps being covered with either plain lanolin or lanolin containing 1000 p.p.m. IAA. The third group of plants was left intact. Three hours after decapitation, labelled sucrose was applied to one of the mature leaves of each plant, and after a further 8 h the apical regions of the plants were assayed for radioactivity. The results (*Table 11.2*) show clearly that the application of IAA to the decapitated plants caused a significant increase in the accumulation of the tracer when compared with those which were treated with lanolin only.

Table 11.2 ACCUMULATION OF METHANOL-SOLUBLE ^{14}C IN TOP INTERNODES OF IAA-TREATED AND DECAPITATED PLANTS OF *Pisum sativem* var. METEOR. (FROM BOOTH *et al.*, 1962, courtesy *Macmillans*)

Treatment	Mean activity (counts/min) of 10 internode samples ± standard error of mean
Intact plant	129.4 ± 27.7
Decapitated plus lanolin	47.7 ± 10.4
Decapitated plus 1000 p.p.m. IAA in lanolin	116.6 ± 25.7

In later work Davies and Wareing (1965) studied the effects of a number of growth substances upon the movement of [32P]phos-phates in pea and cuttings of poplar. In the pea plants IAA was shown to enhance movement of radioactivity towards the decapitated apical parts, this movement taking place in the phloem, as evidenced by steam girdling treatments.

Davies and Wareing then attempted to determine whether the IAA acted locally at the decapitated stump or at a distance from its site of application. They argued that tri-iodobenzoic acid (TIBA), a powerful inhibitor of polar auxin transport, should not have any effect upon IAA-induced phosphate transport if the action of the auxin was merely a local sink effect. On the other hand, if IAA transport was necessary for its action, then application of TIBA should reduce the effect of the IAA. Davies and Wareing, in fact, showed (*Table 11.3*) that TIBA, applied either at the stump with the IAA or between the stump and the site of phosphate application, markedly reduced the accumulation of radioactivity in the stump.

Table 11.3 EFFECT OF TRI-IODOBENZOIC ACID (TIBA) ON IAA-INDUCED TRANSPORT IN PEA STEMS. (FROM DAVIES AND WAREING, 1965, courtesy *Springer-Verlag*)

Treatment	Accumulation of radiophosphorus at the stump (*mean* counts/min ± *standard error*)	
	without TIBA	*with TIBA*
A. TIBA APPLIED TOGETHER WITH IAA TO THE DECAPITATED SURFACE		
Decapitated + IAA	227.7 ± 76.5	16.1 ± 8.6
Decapitated − IAA	3.4 ± 0.9	5.3 ± 0.6
B. TIBA APPLIED BETWEEN THE ^{32}P-SOURCE AND THE STUMP		
Decapitated + IAA	353.2 ± 151.2	31.9 ± 18.6
Decapitated − IAA	11.3 ± 8.5	6.5 ± 2.7

Further work by these investigators, using growth regulators other than IAA on decapitated pea plants revealed that only one, naphthoxyacetic acid, caused a marked increase in ^{32}P-transport to the stump, indole-acetonitrile, 2,4-dichlorophenoxyacetic acid, kinetin and gibberellic acid having no significant effects when compared with lanolin controls.

However, there are reports in the literature which indicate that certain growth regulators other than IAA can have effects upon nutrient transport in leaves. Osborne and Hallaway (1961) showed that butyl esters of 2,4-D cause the retention of ^{14}C-labelled assimilates in detached leaves of *Euonymus*. Mothes and Engelbrecht (1961) demonstrated movement of activity from glycine-1[^{14}C] towards an area treated with kinetin in excised leaves of *Nicotiana*. They suggested that this could be due to an effect of kinetin upon a carrier system responsible for movement. They rejected the idea that movement was due to kinetin-stimulated protein synthesis, i.e. a sink effect, for in another publication (Mothes, Engelbrecht and Schütte, 1961) it was shown that α-aminobutyric acid was subjected to directed transport, although this amino acid was not incorporated into protein. Gunning and Barklay (1963) applied kinetin, gibberellic acid or IAA to oat leaves and determined their effects upon the movement of radioactivity from ^{14}C-labelled glycine or [^{32}P]phosphate. Kinetin caused directed transport towards the hormone-treated area at certain stages of leaf development, gibberellic acid and IAA having no detectable effects.

Seth and Wareing (1967) have reported some interesting effects of mixtures of growth substances, applied to peduncles of *Phaseolus* from which the fruits had been removed, on labelled phosphate movement when this was administered to the lower part of the stem. Although neither kinetin nor gibberellic acid alone had any marked effect on movement, both considerably enhanced the effect of IAA (*Table 11.4*).

Table 11.4 EFFECT OF IAA, KINETIN AND GIBBERELLIC ACID ON ACCUMULATION OF ^{32}P IN PEDUNCLES OF *Phaseolus*. (FROM SETH AND WAREING, 1967, courtesy *The Clarendon Press*)

Treatment	Mean count rate, counts/min	Standard error, $\pm$
Lanolin control	5.9	2.1
Kinetin	19.5	9.6
IAA	234.6	55.1
IAA + kinetin	470.5	100.7
Lanolin control	27.0	2.6
Gibberellic acid	52.9	9.4
IAA	296.7	67.2
IAA + gibberellic acid	676.3	128.8

Although the majority of workers have shown growth substances to enhance nutrient transport towards the site of their application, this does not invariably occur, for in some plants the opposite situation prevails. Hew, Nelson and Krotkov (1967) removed the apical meristems from soyabean plants, the cut surfaces being supplied with IAA, gibberellic acid or water. Half an hour later $^{14}CO_2$ was administered to one of the primary leaves, and the plants were harvested after a further 30 min. It was found that hormone application caused an increase in the total amount of labelled sugars exported from the application leaf as compared with the water controls and also a change in their distribution pattern; more activity went to the roots than to the apical portions after hormone treatment. Hew and his colleagues discounted the possibility of IAA affecting loading from the mesophyll to the sieve elements by showing that no label moved into the mature leaves after application of [^{14}C]IAA to the cut stump.

An opposite view on IAA effects on sugar loading has been expressed by Lepp and Peel (1970), who found an enhanced sucrose efflux from severed aphid stylets sited on bark strips of willow, after application of 10^{-5}M IAA to the cambial surface of the bark. In the same paper experiments with triple-chambered bark strip systems were described: ^{86}RbCl or [^{14}C]sucrose was supplied to the middle compartment, IAA to one of the end compartments, distilled water to the other. The movement of the tracers towards aphid colonies sited on the bark under each end compartment was monitored by assay of the honeydew activities. Labelled rubidium movement appeared to be mainly controlled by sink size, more activity moving towards the largest colony, but sugar movement was influenced markedly by IAA, ^{14}C-activity moving away from the site of IAA application. Kinetin also appeared to have a repulsion effect on labelled sugars.

Apart from the specific issue of hormone effects upon nutrient transport, these experiments of Lepp and Peel are interesting from another point of view. If both the rubidium and the sugars were moving in the sieve tubes, then these experiments could be held to provide good evidence of a certain degree of independence of sugar and cation transport, which may well be worthy of further investigation to produce more meaningful data on this aspect of transport.

Further evidence for a repulsion effect of IAA on labelled sugars has been given by Lepp and Peel (1971b). When tritiated glucose was applied to a bark abrasion situated in the middle of a stem segment of willow, and IAA to an abrasion at one end of the segment, the tritiated sugars moved away from the IAA. In this case longitudinal movement was being investigated, but further work by Lepp and Peel (1971b) on the tangential transport system of willow stems demonstrated that when sugars were moving tangentially around the stem, they moved towards the site of IAA application. Thus, both attraction and repulsion effects of IAA on sugar transport have been demonstrated in the same species.

The situation is even more complicated in willow, for Lepp and Peel (1971b) also showed that when ^{14}C-labelled IAA and ^{3}H-labelled glucose were applied simultaneously at the same site in the middle of a stem segment, they moved to the same degree acropetally and basipetally (*Table 11.5*). Lepp and Peel argued that these results supported the conclusion of Davies and Wareing (1965) concerning a direct effect of IAA upon the longitudinal transport mechanism.

Hormone-directed transport is a most interesting facet of plant physiology which not only concerns those workers who are primarily concerned with the control of growth and differentiation, but may also be of considerable use to those who try to probe the enigma of phloem transport. As mentioned previously, it would be difficult to reconcile certain of the current hypotheses with an unequivocal demonstration that hormones control the direction of nutrient transport via a direct effect upon longitudinal movement. Unfortunately, it is as yet by no means certain that hormones act in such a way.

Table 11.5 PATTERNS OF MOVEMENT OF ^{14}C-LABELLED IAA AND ^{3}H-LABELLED GLUCOSE WHEN BOTH ARE APPLIED SIMULTANEOUSLY TO THE MIDDLE OF A STEM SEGMENT OF WILLOW. (FROM LEPP AND PEEL, 1971b, courtesy *Springer-Verlag*)

| | Experimental stems (IAA+glucose) | | | | Control stems (glucose only) | |
| | ^{14}C counts/min | | ^{3}H counts/min | | ^{3}H counts/min | |
	Apex	Base	Apex	Base	Apex	Base
A. STEM SEGMENTS VERTICAL APEX UPPERMOST						
	260	551	475	999	320	307
	±101	±162	±189	±303	± 83	±104
B. STEM SEGMENTS HORIZONTAL						
	433	392	452	410	270	330
	± 94	±123	± 75	± 83	±108	± 78

Mullins (1970) has produced a comprehensive study of the effects of hormones on [^{14}C]sucrose transport in decapitated plants of *Phaseolus*. Ethylene, IAA, and certain kinins, or mixtures of IAA, gibberellic acid and a kinin, increased accumulation of radioactivity in the cut stump of plants which had been supplied with [^{14}C] sucrose through the epidermis of the petiole of a primary leaf. Abscisic acid decreased transport, this inhibition being relieved by kinins.

Enhanced accumulation of ^{14}C-activity in the treated internodes was apparent 3 h after hormone application. Moreover, Mullins showed that incubation of internodal tissues with hormone mixtures for $2\frac{1}{2}$ h led to an increase in protein and ribonucleic acid synthesis, as evidenced by the enhanced incorporation of [^{14}C]leucine and [^{14}C]orotic acid, respectively. The conclusions drawn by Davies and Wareing (1965) from the results of their experiments using TIBA have also been attacked by Mullins. TIBA at the concentration employed by Davies and Wareing is, Mullins suggested, phytotoxic in *Phaseolus*. Furthermore, Mullins produced data which showed that TIBA, at concentrations which inhibited IAA transport, actually augmented the effect of IAA on sugar accumulation and protein syntheisis in internodal tissues.

The conclusion reached by Mullins that hormones merely exert a sink effect upon the phloem transport system seems at the moment to be reasonable. However, work is proceeding vigorously on this facet of transport physiology, and, no doubt, in the near future publications will appear which throw doubt upon the validity of his inferences.

Concluding remarks

I trust that readers of this book will now have a picture of the difficulties posed to the investigator by the long-distance transport systems of plants, and of the controversies which they engender. The divisions of opinion between transport physiologists are so deep in many areas that I feel that students of the subject are best served if the views of as many workers as possible are presented. Thus, most of the opinions reviewed are not my own, although I have attempted to comment on some of them in the light of my own researches.

Inevitably, it has been a difficult task deciding between the areas which should be emphasised and those which should receive a more cursory treatment; my apologies are extended to those who feel that their efforts have not been justly treated.

Although there is no single piece of evidence which demonstrates unequivocally that sieve tube transport is brought about by a bulk flow of solution, the weight of evidence supports this view. However, as yet it is impossible to decide what mechanisms are responsible for producing this flow; electro-kinetic forces, contractile processes and movement along hydrostatic pressure gradients are all equally strong contenders.

It cannot be emphasised too strongly that the most basic requirement, if we are to solve the problem of the phloem transport mechanism, lies in the elucidation of the structural properties of sieve elements. Until this is done, it seems futile to come down firmly in support of one hypothesis to the exclusion of the others. If P-protein fibrils were shown to be absent from functioning elements, as has been recently suggested by Singh and Srivastava (1972), working on maize phloem, then a major obstacle to pressure flow would be removed and electro-osmosis could be discarded. Similarly, if membrane-bound transcellular strands were observed with the electron microscope, then there seems to be little reason why they should not function in the propulsion of nutrients.

It is not easy to suggest directions in which physiological studies on the mechanism of phloem transport should proceed, since with the available techniques it is difficult to extract meaningful data from the complex phloem system. Work with metabolic inhibitors could prove to be a fruitful field, particularly if means are devised to surmount the problem of their site of action, i.e. whether they inhibit transport mainly by affecting loading processes (Qureshi and Spanner, 1973). Work is also necessary on the detailed metabolism of phloem, the functions of the multitude of enzymes and nucleotides found in sieve tube exudates, and the enzymic and physico-chemical properties of P-protein. Histochemical techniques, particularly those capable of being used in conjunction with the electron microscope, could prove most useful in this field.

As a scientist, one is frequently questioned about the 'usefulness' of pursuing research in a field of 'pure' research such as phloem physiology. Apart from the general answer that one can never predict the outcome of pure research in practical, everyday terms, there are now a number of areas in 'applied' botany which would benefit from a more complete understanding of phloem physiology, one of the most notable of which is pesticide technology.

Although much work has been done on patterns of pesticide movement in plants (Crafts and Crisp, 1971), it is still not known why some pesticides are highly mobile in the phloem or the xylem, while others are not. Since all the available evidence points to the conclusion that selection between different chemical substances occurs during the process of sieve tube loading, it is clear that this is where work on pesticide mobility should be concentrated (in conjunction, of course, with investigations on the metabolic breakdown and immobilisation of the chemical). Techniques are available for studying sieve tube loading; what is required is a great deal of painstaking research.

Transport of nutrients is such an essential aspect of the growth of plants that knowledge of the patterns of movement must play an ever-increasing role in the manipulation of economically important species so that these are able to provide the maximum yield. It would surely be helpful to agronomists if those working on phloem could provide more detailed information as to how the rate and direction of transport are controlled. Once again, what is needed is a vast amount of what might be considered to be unspectacular research, which, however, may finally be of more direct use than considerations of the mechanism of movement.

References

Aikman, D. P. and Anderson, W. P. (1971). 'A Quantitative Investigation of a Peristaltic Model for Phloem Translocation', *Ann. Bot., N.S.*, **35**, 761–772

Alfieri, F. J. and Evert, R. F. (1968). 'Seasonal Development of the Secondary Phloem in *Pinus*', *Am. J. Bot.*, **55**, 518–528

Andel, O. M. van (1953). 'The Influence of Salts on the Exudation of Tomato Plants', *Acta Bot. Neerl.*, **2**, 445–521

Anderson, R. and Cronshaw, J. (1970). 'Sieve Plate Pores in Tobacco and Bean', *Planta (Berl.)*, **91**, 173–180

Anderson, W. P. and Long, J. (1968). 'Longitudinal Water Movement in the Primary Root of *Zea mays*', *J. Exp. Bot.*, **19**, 637–647

Anderssen, F. G. (1929) 'Some Seasonal Changes in the Tracheal Sap of Pear and Apricot Trees', *Plant Physiol. (Lancaster)*, **4**, 459–476

Arisz, W. H. (1945). 'Contribution to a Theory on the Absorption of Salts by the Plant and their Transport in Parenchymatous Tissue', *Proc. K. Ned. Akad. Wet. (Amsterdam)*, **48**, 420–426

Arisz, W. H. (1952). 'Transport of Organic Compounds', *A. Rev. Plant Physiol.*, **3**, 109–130

Arisz, W. H., Helder, R. J. and Nie, R. van (1951). 'Analysis of the Exudation Process in Tomato Plants', *J. Exp. Bot.*, **2**, 257–297

Arnold, W. N. (1968). 'The Selection of Sucrose as the Translocate of Higher Plants', *J. Theor. Biol.*, **21**, 13–20

Avery, G. S. (1935). 'Differential Distribution of a Phytohormone in the Developing Leaf of *Nicotiana*, and its Relation to Polarised Growth', *Bull. Torrey Bot. Soc.*, **62**, 313–330

Barrier, G. E. and Loomis, W. E. (1957). 'Absorption and Translocation of 2,4-Dichlorophenoxyacetic Acid and ^{32}P by Leaves', *Plant Physiol. (Lancaster)*, **32**, 225–231

Bauer. L. (1953). 'Zur Frage der Stoffbewegungen in der Pflanze mit besonderer Berucksichtigung der Wanderung von Fluorochromen', *Planta (Berl.)*, **42**, 367–451

Becker, D., Kluge, M. and Ziegler, H. (1971). 'Der Einbau von ^{32}P—in Organische Verbindungen durch Siebrohrensaft', *Planta (Berl.)*, **99**, 154–162

Behnke, H. D. (1965). 'Über das Phloem der Dioscoreaceen unter besonderer Berücksichtigung inhrer Phloembecken. II. Electronenoptische Untersuchungen zur Feinstruktur des Phloembeckens', *Z. Pflanzenphysiol.*, **53**, 214–244

Behnke, H. D. (1969a). 'Über den Feinbau und die Ausbreitung der Siebröhren-Plasmafilamente und über Bau und Differenzierung der Siebporen bei einigen Monocotylen und bei *Nuphar*', *Protoplasma*, **68**, 377–402

Behnke, H. D. (1969b). 'Über Siebröhren-Plastiden und Plastiden filamente der Caryophyllales', *Planta (Berl.)*, **89**, 275–283

Behnke, H. D. (1971). 'The Contents of the Sieve-Plate Pores in *Aristolochia*', *J. Ultrastruct. Res.*, **36**, 493–498

Behnke, H. D. and Dörr, I. (1967). 'Zur Herkunft und Struktur der Plasmafilamente in Assimilatleitbahnen', *Planta (Berl.)*, **74**, 18–44

Bell, C. W. and Biddulph, O. (1963). 'Translocation of Calcium. Exchange versus Mass Flow', *Plant Physiol. (Lancaster)*, **38**, 610–614

Bennett, C. W. (1937). 'Correlation between Movement of the Curly Top Virus and Translocation of Food in Tobacco and Sugar Beet', *J. Agric. Res.*, **54**, 479–502

Bennett, C. W. (1940a). 'The Relation of Viruses to Plant Tissues', *Bot. Rev.*, **6**, 479–502

Bennett, C. W. (1940b). 'Relation of Food Translocation to the Movement of Virus in Tobacco Mosaic', *J. Agric. Res.*, **60**, 361–390

Biddulph, S. F. (1956). 'Visual Indications of ^{35}S and ^{32}P Translocation in the Phloem', *Am. J. Bot.*, **45**, 648–652

Biddulph, S. F. (1967). 'A Microautoradiographic Study of ^{45}Ca and ^{35}S Distribution in the Intact Bean Root', *Planta (Berl.)*, **74**, 350–367

Biddulph, S. F., Biddulph, O. and Cory, R. (1958). 'Visual Indications of Upward Movement of Foliar Applied ^{32}P and ^{14}C in the Phloem of the Bean Stem', *Am. J. Bot.*, **45**, 648–652

Biddulph, O., Biddulph, S. F., Cory, R. and Koontz, H. (1958). 'Circulation Patterns for Phosphorus, Sulphur and Calcium in the Bean Plant', *Plant Physiol. (Lancaster)*, **33**, 293–300

Biddulph, O. and Cory, R. (1957). 'An Analysis of Translocation in the Phloem of the Bean Plant Using THO, ^{32}P and ^{14}CO$_2$', *Plant Physiol. (Lancaster)*, **32**, 608–619

Biddulph, O. and Cory, R. (1960). 'Demonstration of Two Translocation Mechanisms in Studies of Bidirectional Movement', *Plant Physiol. (Lancaster)*, **35**, 689–695

Biddulph, O. and Cory, R. (1965). 'Translocation of ^{14}C-Metabolites in the Phloem of the Bean Plant', *Plant Physiol. (Lancaster)*, **40**, 119–129

Biddulph, O. and Markle, J. (1944). 'Translocation of Radiophosophorus in the Phloem of the Cotton Plant', *Am. J. Bot.*, **31**, 65–70

Bieleski, R. L. (1962). 'The Physiology of Sugar Cane *v*. Kinetics of Sugar Accumulation', *Aust. J. Biol. Sci.*, **15**, 429–444

Bieleski, R. L. (1966). 'Accumulation of Phosphate, Sulphate and Sucrose by Excised Phloem Tissues', *Plant Physiol. (Lancaster)*, **41**, 447–454

Bieleski, R. L. (1969). 'Phosphorus Compounds in Translocating Phloem', *Plant Physiol. (Lancaster)*, **44**, 497–502

Birch-Hirschfield, L. (1920) 'Untersuchungen über die Ausbreitungsgeschwindigkeit gelöster Stoffe in der Pflanze', *Jb. Wiss Bot.*, **59**, 171–262

Black, M. K. and Osborne, D. J. (1965). 'Polarity of Transport of Benzyladenine, Adenine and Indole-3-acetic Acid in Petiole Segments of *Phaseolus vulgaris*', *Plant Physiol. (Lancaster)*, **40**, 676–680

Böhning, R. H., Kendall, W. A. and Linck, A J (1953). 'The Effect of Temperature on Growth and Translocation in the Tomato', *Am. J. Bot.*, **40**, 150–153

Bollard, E. G. (1953). 'The Use of Tracheal Sap in the Study of Apple Tree Nutrition', *J. Exp. Bot.*, **4**, 363–368

Bollard, E. G. (1956). 'Nitrogenous Compounds in Plant Xylem Sap', *Nature (Lond.)*, **178**, 1189–1190

Bollard, E. G. (1957a). 'Composition of the Nitrogen Fraction of Apple Tracheal Sap', *Aust. J. Biol. Sci.*, **10**, 279–287

Bollard, E. G. (1957b). 'Nitrogenous Compounds in Tracheal Sap of Woody Members of the Family Rosaceae', *Aust. J. Biol. Sci.*, **10**, 288–291

Bollard, E. G. (1960). 'Transport in the Xylem', *A. Rev. Plant Physiol.*, **11**, 141–166

Bonnemain, J. L. (1970). 'Transport de l'AIA Marqué et de ses Dérivés à Partir des Jeunes Fruits', *C.R. Hebd. Séanc. Acad. Sci. Paris*, **270**, 1326–1329

Booth, A., Moorby, J., Davies, C. R., Jones, H. and Wareing, P. F. (1962). 'Effects of Indolyl-3-acetic Acid on the Movement of Nutrients in Plants', *Nature (Lond.)*, **194**, 204–205

Bouck, G. B. and Cronshaw, J. (1965). 'The Fine Structure of Differentiating Sieve Tube Elements', *J. Cell. Biol.*, **25**, 79–96

Bowen, M. R. and Hoad, G. V. (1968). 'Inhibitor Content of Phloem and Xylem Sap Obtained from Willow (*Salix viminalis* L.) Entering Dormancy', *Planta (Berl.)*, **81**, 64–70

Bowen, M. R. and Wareing, P. F. (1969). 'The Interchange of ^{14}C-Kinetin and ^{14}C-Gibberellic Acid between Bark and Xylem of Willow', *Planta (Berl.)*, **89**, 108–125

Bowling, D. J. F. (1968a). 'Measurement of the Potential Across the Sieve Plates in *Vitis vinifera*', *Planta (Berl.)*, **80**, 21–26

Bowling, D. J. F. (1968b). 'Translocation at 0°C in *Helianthus annuus*', *J. Exp. Bot.*, **19**, 381–388

Bowling, D. J. F. (1969). 'Evidence for the Electro-osmosis Theory of Transport in the Phloem', *Biochem. Biophys. Acta.*, **183**, 230–232

Bowling, D. J. F. and Spanswick, R. M. (1964). 'Active Transport of Ions across the Root Cortex of *Ricinus communis*', *J. exp. Bot.*, **15**, 422–427

Bowling, D. J. F. and Weatherley, P. E. (1965). 'The Relationships between Transpiration and Potassium Uptake in *Ricinus communis*'. *J. Exp. Bot.*, **16**, 732–741

Brady, H. A. (1969). 'Light Intensity and the Absorption and Translocation of 2,4,5-T by Woody Plants', *Weed Sci.*, **17**, 320–322

Briggs, G. E. (1957), 'Some Aspects of Free Space in Plant Tissues', *New Phytol.*, **56**, 305–324

Briggs, G. E., Hope, A. B. and Pitman, M. G. (1958). 'Exchangeable Ions in Beet Disks at Low Temperature', *J. Exp. Bot.*, **9**, 128–141

Brouwer, R. (1954). 'The Regulating Influence of Transpiration and Suction Tension on the Water and Salt Uptake by Roots of Intact *Vicia faba* Plants, *Acta. Bot. Neerl.*, **3**, 264–312

Brouwer, R. (1965). 'Ion Absorption and Transport in Plants', *A. Rev. Plant Physiol.*, **16**, 241–266

Broyer, T. C. and Hoagland, D. R. (1943). 'Metabolic Activities of Roots and their Bearing on the Relation of Upward Movement of Salts and Water in Plants', *Am. J. Bot.*, **30**, 261–273

Bryant, A. E. (1934). 'A Demonstration of the Connection of the Protoplasts of the Endodermal Cells with the Casparian Strips in the Roots of Barley', *New Phytol.*, **33**, 231

Canning, R. E. and Kramer, P. J. (1958). 'Salt Absorption and Accumulation in Various Regions of Roots', *Am. J. Bot.*, **45**, 378–382

Canny, M. J. (1960a). 'The Breakdown of Sucrose during Translocation', *Ann. Bot.*, *N.S.*, **24**, 330–344

Canny, M. J. (1960b). 'The Rate of Translocation', *Biol. Rev.*, **35**, 507–532

Canny, M. J. (1961). 'Measurements of the Velocity of Translocation', *Ann. Bot.*, *N.S.*, **25**, 152–167

Canny, M. J. (1962). 'The Mechanism of Translocation', *Ann. Bot.*, *N.S.*, **26**, 603–617

Canny, M. J. (1971). 'Translocation: Mechanisms and Kinetics', *A. Rev. Plant Physiol.*, **22**, 237–260

Canny, M. J. and Askham, M. J. (1967). 'Physiological Inferences from the Evidence of Translocated Tracer: A Caution', *Ann. Bot.*, *N.S.*, **31**, 409–416

Canny, M. J. and Markus, K. (1960). 'The Metabolism of Phloem Isolated from Grapevine', *Aust. J. Biol. Sci.*, **13**, 292–299

Canny, M. J., Nairn, B. and Harvey, N. (1968). 'The Velocity of Translocation in Trees', *Aust. J. Bot.*, **16**, 479–485

Canny, M. J. and Phillips, O. M. (1963). 'Quantitiative Aspects of a Theory of Translocation', *Ann. Bot.*, *N.S.*, **27**, 379–402

Carr, D. J., Reid, D. M. and Skene, K. G. M. (1964). 'The Supply of Gibberellins from the Root to the Shoot', *Planta (Berl.)*, **63**, 382–392

Carr, D. J. and Wardlaw, I. F. (1965). 'The Supply of Photosynthetic Assimilates to the Grain from the Flag Leaf and Ear of Wheat', *Aust. J. Biol. Sci.*, **18**, 711–719

Cataldo, D. A., Christy, A. L. and Coulson, C. L. (1972). 'Solution Flow in Phloem. II. Phloem Transport of THO in *Beta vulgaris*', *Plant Physiol. (Lancaster)*, **49**, 690–695

Cataldo, D. A., Christy, A. L., Coulson, C. L. and Ferrier, J. M. (1972). 'Solution Flow in Phloem. I. Theoretical Considerations', *Plant Physiol. (Lancaster)*, **49**, 685–689

Charles, A. (1953). 'Uptake of Dyes into Cut Leaves', *Nature (Lond.)*, **171**, 435–436

Chen, S. L. (1951). 'Simultaneous Movement of ^{32}P and ^{14}C in Opposite Directions in Phloem Tissue', *Am. J. Bot.*, **38**, 203–211

Choi, I. C. and Aronoff, S. (1966). 'Photosynthate Transport Using Tritiated Water', *Plant Physiol. (Lancaster)*, **41**, 1119–1129

Clements, H. F. (1940). 'Movement of Organic Solutes in the Sausage Tree, *Kigelia africana*', *Plant Physiol. (Lancaster)*, **15**, 689–700

Clor, M. A. (1967). 'Translocation of Tritium-labelled Gibberellic Acid in Pea Stem Segments or Potato Tuber Cylinders', *Nature (Lond.)*, **214**, 1263–1264

Colwell, R. N. (1942). 'The Use of Radioactive Phosphorus in Translocation Studies', *Am. J. Bot.*, **29**, 798–807

Cooper, W. C. (1936). 'Transport of Root-Forming Hormone in Woody Cuttings', *Plant Physiol. (Lancaster)*, **11**, 779–793

Cormack, R. G. H. and Lemay, P. (1963). 'Sugar in the Intercellular Spaces of White Mustard Roots', *J. Exp. Bot.*, **14**, 232–236

Coulson, C. L. and Peel, A. J. (1968). 'Respiration of ^{14}C-labelled Assimilates in Stems of Willow', *Ann. Bot. N.S.*, **32**, 867–876

Coulson, C. L. and Peel, A. J. (1971). 'The Effect of Temperature on the Respiration of ^{14}C-labelled Sugars in Stems of Willow', *Ann. Bot.*, *N.S.*, **35**, 9–15

Coupland, D. and Peel, A. J. (1972). 'Maleic Hydrazide as an Antimetabolite of Uracil', *Planta (Berl.)*, **103**, 249–253

Crafts, A. S. (1931). 'Movement of Organic Materials in Plants', *Plant Physiol. (Lancaster)*, **6**, 1–41

Crafts, A. S. (1932). 'Phloem Anatomy, Exudation and Transport of Organic Nutrients in Cucurbits', *Plant Physiol. (Lancaster)*, **7**, 183–225

Crafts, A. S. (1933). 'Sieve Tube Structure and Translocation in the Potato', *Plant Physiol. (Lancaster)*, **8**, 81–104

Crafts, A. S. (1936). 'Further Studies on Exudation in Curcurbits', *Plant Physiol.* (*Lancaster*), **11**, 63–79

Crafts, A. S. (1951). 'Movement of Assimilates, Viruses, Growth Regulators and Chemical Indicators in Plants', *Bot. Rev.*, **17**, 203–284

Crafts, A. S. and Broyer, T. C. (1938). 'Migration of Salts and Water into the Xylem of the Roots of Higher Plants', *Am. J. Bot.*, **25**, 529–535

Crafts, A. S. and Crisp, C. E. (1971). *Phloem Transport in Plants.* Freeman, San Francisco

Crafts, A. S. and Lorenz, O. A. (1944). 'Fruit Growth and Food Transport in Cucurbits', *Plant Physiol.* (*Lancaster*), **19**, 131–138

Cronshaw, J. (1969). '*Nicotiana* Sieve Plate Pores', *Proc. Calif. Soc. Electronmicrosc.*, **4**, A5

Cronshaw, J. (1974). 'Phloem Differentiation and Development', in *Dynamic Aspects of Plant Ultrastructure* (Ed. A. W. Robards). McGraw-Hill Book Company (UK) Ltd

Cronshaw, J. and Anderson, R. (1969). 'Sieve Plate Pores of *Nicotiana*', *J. Ultrastruct. Res.*, **27**, 134–148

Cronshaw, J. and Esau, K. (1967). 'Tubular and Fibrillar Components of Mature and Differentiating Sieve Elements', *J. Cell. Biol.*, **34**, 801–816

Cronshaw, J. and Esau, K. (1968a). 'P-Protein in the Phloem of *Cucurbita*. I. The Development of P-Protein Bodies', *J. Cell. Biol.*, **38**, 25–39

Cronshaw, J. and Esau, K. (1968b) 'P-Protein in the Phloem of *Cucurbita*. II. The P-Protein of Mature Sieve Elements', *J. Cell. Biol.*, **38**, 292–303

Crossett, R. N. (1967). 'Autoradiography of ^{32}P in Maize Roots', *Nature* (*Lond.*), **213**, 312–313

Crowdy, S. H. and Tanton, T. W. (1970). 'Water Pathways in Higher Plants. I. Free Space in Wheat Leaves', *J. Exp. Bot.*, **21**, 102–111

Currier, H. B., Esau, K. and Cheadle, V. I. (1955). 'Plasmolytic Studies of Phloem', *Am. J. Bot.*, **42**, 68–81

Currier, H. B., McNairn, R. B. and Webster, D. H. (1966). 'Blockage Phenomena in Axial Phloem Conduction', *Plant Physiol.* (*Lancaster*), **41** (Supplement), XX

Curtis, O. F. (1929). 'Studies on Solute Translocation in Plants. Experiments Indicating that Translocation is Dependent on the Activity of Living Cells', *Am. J. Bot.*, **16**, 154–168

Curtis, O. F. (1935). *The Translocation of Solutes in Plants.* McGraw-Hill, New York

Curtis, O. F. and Herty, S. D. (1936). 'The Effect of Temperature on Translocation from Leaves', *Am. J. Bot.*, **23**, 528–532

Curtis, O. F. and Schofield, H. T. (1933). 'A Comparison of Osmotic Concentrations of Supplying and Receiving Tissues and its Bearing on the Münch Hypothesis of the Translocation Mechanism', *Am. J. Bot.*, **20**, 502–512

Dainty, J. (1962). 'Ion Transport and Electrical Potentials in Plant Cells', *A. Rev. Plant Physiol.*, **13**, 379–402

Dainty, J. and Hope, A. B. (1959). 'Ionic Relations of Cells of *Chara australis*. I. Ion Exchange in the Cell Wall', *Aust. J. Biol. Sci.*, **12**, 395–411

Davies, C. R. and Wareing, P. F. (1965). 'Auxin-directed Transport of Radiophosphorus in Stems', *Planta* (*Berl.*), **65**, 139–156

Davis, J. D. and Evert, R. F. (1970). 'Seasonal Cycle of Development in Woody Vines', *Bot. Gaz.*, **131**, 128–138

De la Fuente, R. K. and Leopold, A. C. (1966). 'Kinetics of Polar Auxin Transport', *Plant Physiol.* (*Lancaster*), **41**, 1481–1484

De Vries, H. (1885). 'Über die Bedeutung der Circulation und der Rotation des Protoplasma für das Stofftransport in der Pflanze', *Bot. Ztg.*, **43**, 1–26

Die, J. van and Tammes, P. M. L. (1966). 'Studies on Phloem Exudation from *Yucca flaccida* Haw. III. Prolonged Bleeding from Isolated Parts of the Young

Inflorescence', *Proc. K. Ned. Akad. Wet. (Amsterdam)*, **69**, 648–654

Dixon, H. H. (1914). *Transpiration and the Ascent of Sap in Plants*. Macmillan, London

Dixon, H. H. and Ball, N. G. (1922). 'Transport of Organic Substances in Plants', *Nature (Lond.)*, **109**, 236–237

Doi, Y., Teranaka, M., Yora, K. and Asuyama, H. (1967). 'Mycoplasma or PLT Group-like Micro-organisms Found in the Phloem Elements of Plants Infected with Mulberry Dwarf, Potato Witch's Broom, Aster Yellows or Pauwlonia Witch's Broom', *Ann. Phytopath. Soc. Japan*, **33**, 259–266

Duloy, M. D. and Mercer, F. V. (1961). 'Studies in Translocation. I. The Respiration of the Phloem', *Aust. J. Biol. Sci.*, **14**, 391–401

Duloy, M., Mercer, F. V. and Rathgeber, N. (1961). 'Studies in Translocation. II. Submicroscopic Anatomy of the Phloem', *Aust. J. Biol. Sci.*, **14**, 506–518

Duloy, M., Mercer, F. V. and Rathgeber, N. (1962). 'Studies in Translocation. III. The Cytophysiology of the Phloem of *Cucurbita pepo*', *Aust. J. Biol. Sci.*, **15**, 459–467

Dunlop, J. and Bowling, D. J. F. (1971). 'The Movement of Ions to the Xylem Exudate of Maize Roots. I. Profiles of Membrane Potential and Vacuolar Potassium Activity Across the Root', *J. Exp. Bot.*, **22**, 434–444

Eaton, F. M. and Ergle, D. R. (1948). 'Carbohydrate Accumulation in the Cotton Plant at Low Moisture Levels', *Plant Physiol. (Lancaster)*, **23**, 169–187

Ehara, K. and Sekioka, H. (1962). 'Effect of Atmospheric Humidity and Soil Moisture on the Translocation of Sucrose-^{14}C in the Sweet Potato Plant', *Proc. Crop Sci. Soc. Japan*, **31**, 41–44

Eliasson, L. (1965). 'Interference of the Transpiration Stream with the Basipetal Translocation of Leaf Applied Chlorophenoxy Herbicides in Aspen (*Populus tremula* L.), *Physiologia Plant.*, **18**, 506–515

Engleman, E. M. (1965a). 'Sieve Elements of *Impatiens sultanii*. I. Wound Reaction', *Ann. Bot., N.S.*, **29**, 83–101

Engleman, E. M. (1956b) 'Sieve Elements of *Impatiens sultanii*. II. Developmental Aspects', *Ann. Bot., N.S.*, **29**, 103–118

Esau, K. (1948). 'Phloem Structure in the Grapevine and its Seasonal Changes', *Hilgardia*, **18**, 217–296

Esau, K. (1965). *Vascular Differentiation in Plants*. Holt, Rinehart and Winston, New York

Esau, K. (1967). 'Anatomy of Plant Virus Infections', *A. Rev. Phytopathol.*, **5**, 45–76

Esau, K. (1969). 'The Phloem', in *Handbuch der Pflanzen Anatomie*, Bd. 5, T.2. Borntraeger, Berlin

Esau, K. and Cheadle, V. I. (1961). 'An Evaluation of Studies on the Ultrastructure of Sieve Plates', *Proc. Natn. Acad. Sci., U.S.A.*, **47**, 1716–1726

Esau, K. and Cheadle, V. I. (1962a). 'Mitochondria in the Phloem of *Cucurbita*', *Bot. Gaz.*, **124**, 79–85

Esau, K. and Cheadle, V. I. (1962b). 'An Evaluation of Studies on Ultrastructure of Tonoplast in Sieve Elements', *Proc. Natn. Acad Sci., U.S.A.*, **46**, 1–8

Esau, K. and Cheadle, V. I. (1965) 'Cytologic Studies on Phloem', *Univ. Calif. Publs. Bot.*, **36**, 253–344

Esau, K., Cheadle, V. I. and Risley, E. B. (1962). 'Development of Sieve-plate Pores', *Bot. Gaz.*, **123**, 233–243

Esau K. and Cronshaw, J. (1968a). 'Plastids and Mitochondria in the Phloem of *Cucurbita*', *Can. J. Bot.*, **46**, 877–880

Esau, K. and Cronshaw, J. (1968b). 'Endoplasmic Reticulum in the Sieve Element of *Cucurbita*', *J. Ultrastruct. Res.*, **23**, 1–14

Esau, K., Cronshaw, J. and Hoefert, L. L. (1968). 'Relation of Beet Yellows Virus to the Phloem and to Movement in the Sieve Tube', *J. Cell. Biol.*, **32**, 71–87

Esau, K., Engleman, E. M. and Bisalputra, T. (1963). 'What are Transcellular Strands?' *Planta (Berl.)*, **59**, 617–623

Eschrich, W. (1963). 'Beziehungen Zwischen dem Auftreten von Callose und der Feinstruktur des Primären Phloems bei *Cucurbita ficifolia*', *Planta (Berl.)*, **59**, 243–261

Eschrich, W. (1967). 'Bidirektionelle Translokation in Siebröhren', *Planta (Berl.)*, **73**, 37–49

Eschrich, W. (1968). 'Translokation Radioactiv Markierter Indolyl-3-Essigsäure in Siebröhren von *Vicia faba*', *Planta (Berl.)*, **78**, 147–157

Eschrich, W. (1970). 'Biochemistry and the Fine Structure of Phloem in Relation to Transport', *A. Rev. Plant Physiol.*, **21**, 193–214

Eschrich, W., Evert, R. F. and Heyser, W. (1971). 'Proteins of the Sieve Tube Exudate of *Cucurbita maxima*', *Planta (Berl.)*, **100**, 208–221

Eschrich, W., Yamaguchi, S. and McNairn, R. B. (1965). 'Der Einfluss verstärkter Callosebildung auf den Stofftransport in Siebröhren', *Planta (Berl.)*, **65**, 49–64

Evans, N. T. S., Ebert, M. and Moorby, J. (1963). 'A Model for the Translocation of Photosynthate in the Soybean', *J. Exp. Bot.*, **14**, 221–231

Evert, R. F. (1962). 'Some Aspects of Phloem Development in *Tilia americana*', *Am. J. Bot.*, **49**, 659

Evert, R. F. and Alfieri, F. J. (1965). 'Ontogeny and Structure of Coniferous Sieve Cells', *Am. J. Bot.*, **52**, 1058–1066

Evert, R. F., Davis, J. D., Tucker, C. M. and Alfieri, F. J. (1970). 'On the Occurrence of Nuclei in Mature Sieve Elements', *Planta (Berl.)*, **95**, 281–296

Evert, R. F. and Derr, W. E. (1964). 'Slime Substance and Strands in Sieve Elements', *Am. J. Bot.*, **51**, 875–880

Evert, R. F. and Deshpande, B. P. (1970). 'Nuclear P-Protein in Sieve Elements of *Tilia americana*', *J. Cell. Biol.*, **44**, 462–466

Evert, R. F. and Deshpande, B. P. (1971). 'Plastids in Sieve Elements and Companion Cells of *Tilia americana*', *Planta (Berl.)*, **96**, 97–100

Evert, R. F., Eschrich, W. and Eichorn, S. E. (1971). 'Sieve Plate Pores in Leaf Veins of *Hordeum vulgare*', *Planta. (Berl.)*, **100**, 262–267

Evert, R. F., Eschrich, W., Medler, J. T. and Alfieri, F. J. (1968). 'Observations on the Penetration of Linden Branches by Stylets of the Aphid *Longistigma caryae*', *Am. J. Bot.*, **55**, 860–874

Evert, R. F. and Murmanis, L. (1965). 'Ultrastructure of the Secondary Phloem of *Tilia americana*', *Am. J. Bot.*, **52**, 95–106

Evert, R. F., Murmanis, L. and Sachs, T. B. (1966). 'Another View of the Ultrastructure of *Cucurbita* Phloem', *Ann. Bot., N.S.*, **30**, 563–585

Evert, R. F., Tucker, C. M., Davis, J. D. and Deshpande, B. P. (1969). 'Light Microscope Investigation of Sieve Element Ontogeny and Structure in *Ulmus americana*', *Am. J. Bot.*, **56**, 999–1017

Fensom, D. S. (1957). 'The Bio-electric Potentials of Plants and their Functional Significance. I. An Electro-kinetic Theory of Transport', *Can. J. Bot.*, **35**, 573–582

Fensom, D. S. (1972). 'A Theory of Translocation in Phloem of *Heracleum* by Contractile Protein Microfibrillar Material', *Can. J. Bot.*, **50**, 479–497

Fensom, D. S. and Davidson, H. R. (1970). 'Microinjection of ^{14}C-Sucrose into Single Living Sieve Tubes of *Heracleum*', *Nature (Lond.)*, **227**, 857–858

Fensom, D. S. and Spanner, D. C. (1969). 'Electro-osmotic and Biopotential Measurements of Phloem Strands of *Nymphoides*', *Planta (Berl.)*, **88**, 321–331

Field, R. J. and Peel, A. J. (1971a). 'The Metabolism and Radial Movement of Growth Regulators and Herbicides in Willow Stems', *New Phytol.*, **70**, 743–749

Field, R. J. and Peel, A. J. (1971b). 'The Movement of Growth Regulators and Herbicides into the Sieve Elements of Willow', *New Phytol.*, **70**, 997–1003

Field, R. J. and Peel, A. J. (1972). 'The Longitudinal Mobility of Growth Regulators and Herbicides in Sieve Tubes of Willow', *New Phytol.*, **71**, 249–254

Fletcher, R. A. and Zalik, S. (1965). 'Effects of Light of Several Spectral Bands on the Metabolism of Radioactive IAA in Bean Seedlings', *Plant Physiol. (Lancaster)*, **40**, 549–552

Ford, J. (1967). 'Studies on Lateral and Longitudinal Transport in the Sieve Tubes of Higher Plants', Ph.D. Thesis, University of Hull

Ford, J. and Peel, A. J. (1966). 'The Contributory Length of Sieve Tubes in Isolated Segments of Willow and the Effect on It of Low Temperatures', *J. Exp. Bot.*, **17**, 522–533

Ford, J. and Peel, A. J. (1967a). 'Preliminary Experiments on the Effect of Temperature on the Movement of ^{14}C-Labelled Assimilates through the Phloem of Willow', *J. Exp. Bot.*, **18**, 406–415

Ford, J. and Peel, A. J. (1967b). 'The Movement of Sugars into the Sieve Elements of Bark Strips of Willow. I. Metabolism During Transport', *J. Exp. Bot.*, **18**, 607–619

Forde, B. J. (1966). 'Translocation in Grasses. I. Bermuda Grass', *New Zealand J. Bot.*, **4**, 479–495

Fritz, E. and Eschrich, W. (1970). '^{14}C-Mikroautoradiographie Wasserlöslicher Substanzen im Phloem', *Planta (Berl.)*, **92**, 267–281

Fujiwara, A and Suzuki, M. (1961). 'Effects of Temperature and Light on the Translocation of Photosynthetic Products', *Tohoku J. Agric. Res.*, **12**, 363–367

Funderburk, H. H. and Lawrence, J. M. (1964). 'Mode of Action and Metabolism of Diquat and Paraquat', *Weeds*, **12**, 259–264

Gage, R. S. and Aronoff, S. (1960). 'Translocation. III. Experiments with Carbon-14, Chlorine-36 and Hydrogen-3', *Plant Physiol. (Lancaster)*, **35**, 53–64

Gardner, D. C. J. and Peel, A. J. (1969). 'ATP in Sieve Tube Sap from Willow', *Nature (Lond.)*, **222**, 774

Gardner, D. C. J. and Peel, A. J. (1971). 'Transport of Sugars into the Sieve Elements of Willow', *Phytochemistry*, **10**, 2621–2625

Gardner, D. C. J. and Peel, A. J. (1972a). 'The Effect of Low Temperature on Sucrose, ATP and Potassium Concentrations and Fluxes in the Sieve Tubes of Willow', *Planta (Berl.)*, **102**, 348–356

Gardner, D. C. J. and Peel, A. J. (1972b). 'Some Observations on the Role of ATP in Sieve Tube Translocation', *Planta (Berl.)*, **107**, 217–226

Geiger, D. R. (1966). 'Effect of Sink Region Cooling on Translocation of Photosynthate', *Plant Physiol. (Lancaster)*, **41**, 1667–1672

Geiger, D. R. (1969). 'Chilling and Translocation Inhibition', *Ohio J. Sci.*, **69**, 356–366

Geiger, D. R. and Batey, J. W. (1967). 'Translocation of ^{14}C-Sucrose in Sugar Beet During Darkness', *Plant Physiol. (Lancaster)*, **42**, 1743–1749

Geiger, D. R. and Christy, A. L. (1971). 'Effect of Sink Region Anoxia on Translocation Rate', *Plant Physiol. (Lancaster)*, **47**, 172–174

Geiger, D. R. and Swanson, C. A. (1965a). 'Sucrose Translocation in Sugar Beet', *Plant Physiol. (Lancaster)*, **40**, 685–690

Geieger, D. R. and Swanson, C. A. (1965b). 'Evaluation of Selected Parameters in a Sugar Beet Translocation System', *Plant Physiol. (Lancaster)*, **40**, 942–947

Giannotti, J., Devauchelle, G. and Vago, C. (1970). 'Recherches sur le Cycle de Developpement des Mycoplasmes de Plantes Transmis par des Vecteurs',

Congr. Int. Microsc. Electron., Grenoble, **III,** 353-354

Goldsmith, M. H. M. (1966). 'Movement of Indoleacetic Acid in Coleoptiles of *Avena sativa* L. II. Suspension of Polarity by Total Inhibition of the Basipetal Transport', *Plant. Physiol. (Lancaster)*, **41,** 15–27

Goldsmith, M. H. M. (1968). 'The Transport of Auxin', *A. Rev. Plant Physiol.*, **19,** 347–360

Goldsmith, M. H. M. (1969), 'Transport of Plant Growth Regulators', in *The Physiology of Plant Growth and Development* (Ed. M. B. Wilkins). McGraw-Hill Book Company (UK) Ltd

Gorter, C. J. and Veen, H. (1966). 'Auxin Transport in Explants of *Coleus*', *Plant Physiol. (Lancaster)*, **41,** 83–86

Greenidge, K. N. H. (1957). 'Sap Ascent in Trees', *A. Rev. Plant Physiol.*, **8,** 237–256

Gunning, B. E. S. and Barklay, W. K. (1963). 'Kinin-induced Directed Transport and Senescence in Detached Oat Leaves', *Nature (Lond.)*, **199,** 262–265

Gunning, B. E. S. and Pate, J. S. (1969). ' "Transfer Cells". Plant Cells with Wall Ingrowths, Specialised in Relation to Short Distance Transport of Solutes—Their Occurrence, Structure, and Development', *Protoplasma*, **68,** 107–133

Gunning, B. E. S. and Pate, J. S. (1974). 'Transfer Cells', in *Dynamic Aspects of Plant Ultrastructure* (Ed. A. W. Robards). McGraw-Hill Book Company (UK) Ltd

Hales, S. (1961). *Vegetable Staticks*, 2nd edn. Oldbourne, London

Hall, S. M., Baker, D. A. and Milburn, J. A. (1971). 'Phloem Transport of ^{14}C-labelled Assimilates in *Ricinus*', *Planta (Berl.)*, **100,** 200–207

Hammel, H. T. (1968). 'Measurement of Turgor Pressure and its Gradient in the Phloem of Oak'. *Plant Physiol. (Lancaster)*, **43,** 1042–1048

Happel, J. and Brenner, H. (1965). *Low Reynolds Number Hydrodynamics: With Special Applications to Particulate Media*, 1st edn. Prentice-Hall, Englewood Cliffs, N.J.

Harel, S. and Reinhold, L. (1966). 'The Effect of 2,4-Dinitrophenol on Translocation in the Phloem', *Physiologia Plant.*, **19,** 634–643

Hartig, T. (1837). 'Jahresberichte über die Fortschritte der Forstwissenschaften und forstlichen', *Naturkunde*, **1,** 125–168

Hartig, T. (1861). 'Ueber die Bewegung des Saftes in den Holzpflanzen', *Bot. Ztg.*, **19,** 17–23

Hartig, T. (1877). *Lehrbuch für Förster*, 11th edn. J. G. Cotta, Stuttgart

Hartt, C. E. (1963). 'Translocation as a Factor in Photosynthesis', *Naturwissenschaften*, **50,** 666–667

Hartt, C. E. (1965a). 'The Effect of Temperature upon Translocation of ^{14}C in Sugar Cane', *Plant Physiol. (Lancaster)*, **40,** 74–81

Hartt, C. E. (1965b). 'Light and Translocation of ^{14}C in Detached Blades of Sugarcane', *Plant Physiol. (Lancaster)*, **40,** 718–724

Hartt, C. E. (1966). 'Translocation in Coloured Light', *Plant Physiol. (Lancaster)*, **41,** 369–372

Hartt, C. E. (1967). 'Effect of Moisture Supply on Translocation and Storage of ^{14}C in Sugar Cane', *Plant Physiol. (Lancaster)*, **42,** 338–346

Hartt, C. E., Kortschak, H. P. and Burr, G. O. (1964). 'Effects of Defoliation, Deradication and Darkening the Blade upon Translocation of ^{14}C in Sugar Cane'. *Plant Physiol. (Lancaster)*, **39,** 15–22

Hatch, M. D. (1964). 'Sugar Accumulation by Sugar Cane Storage Tissue: the Role of Sucrose Phosphate', *Biochem. J.*, **93,** 521–526

Hatch, M. D. and Glasziou, K. T. (1963). 'Sugar Accumulation Cycle in Sugar

Cane. II. Relationship of Invertase Activity to Sugar Content and Growth Rate in Storage Tissue of Plants Grown in Controlled Environments', *Plant Physiol. (Lancaster)*, **38**, 344–348

Hatch, M. D. and Glasziou, K. T. (1964). 'Direct Evidence for Translocation of Sucrose in Sugarcane Leaves and Stems', *Plant Physiol. (Lancaster)*, **39**, 180–184

Hawker, J. S. (1965). 'The Sugar Content of Cell Walls and Intercellular Spaces in Sugar Cane Stems and its Relation to Sugar Transport', *Aust. J. Biol. Sci.*, **18**, 959–969

Hepton, C. E. L. and Preston, R. D. (1960). 'Electron Microscopic Observations of the Structure of Sieve-connections in the Phloem of Angiosperms and Gymnosperms', *J. Exp. Bot.*, **11**, 381–394

Hepton, C. E. L., Preston, R. D. and Ripley, G. W. (1955). 'Electron Microscopic Observations on the Structure of the Sieve Plates in *Cucurbita*', *Nature (Lond.)*, **176**, 868–870

Hertel, R. and Leopold, A. C. (1963). 'Versuche zur Analyse des Auxintransportes in der Koleoptile von *Zea mays* L.', *Planta (Berl.)*, **59**, 535–562

Hew, C. S., Nelson, C. D. and Krotkov, G. (1967), 'Hormonal Control of Translocation of Photosynthetically Assimilated ^{14}C in Young Soybean Plants', *Am. J. Bot.*, **54**, 252–256

Hewitt, S. P. and Curtis, O. F. (1948). 'The Effect of Temperature on Loss of Dry Matter and Carbohydrate from Leaves by Respiration and Translocation', *Am. J. Bot.*, **35**, 746–755

Heyser, W., Eschrich, W. and Evert, R. F. (1969). 'Translocation in Perennial Monocotyledons', *Science, N.Y.*, **164**, 572–574

Ho, L. C. and Mortimer, D C (1971). 'The Site of Cyanide Inhibition of Sugar Translocation in Sugar Beet Leaf', *Can. J. Bot.*, **49**, 1769–1775

Ho, L. C. and Peel, A. J. (1969a). 'Transport of ^{14}C-labelled Assimilates and ^{32}P-labelled Phosphate in *Salix viminalis* in Relation to Phyllotaxis and Leaf Age', *Ann. Bot., N.S.*, **33**, 743–751 .

Ho, L. C. and Peel, A. J. (1969b). 'Investigation of Bidirectional Movement of Tracers in Sieve Tubes of *Salix viminalis* L.', *Ann. Bot., N.S.*, **33**, 833–844

Hoad, G. V. (1967). '(+)-Abscisin II, ((+)-Dormin) in Phloem Exudate of Willow', *Life Sci.*, **6**, 1113–1118

Hoad, G. V. and Bowen, M. R. (1968). 'Evidence for Gibberellin-like Substances in the Phloem Exudate of Higher Plants', *Planta (Berl.)*, **82**, 22–32

Hoad, G. V., Hillman, S. K. and Wareing, P. F. (1971). 'Studies on the Movement of Indole Auxins in Willow (*Salix viminalis* L.)', *Planta (Berl.)*, **99**, 73–88

Hoad, G. V. and Peel. A. J. (1965a). 'Studies on the Movement of Solutes between the Sieve Tubes and Surrounding Tissues in Willow. I. Interference between Solutes and Rate of Translocation Measurements', *J. Exp. Bot.*, **16**, 433–451

Hoad, G. V. and Peel, A. J. (1965b). 'Studies on the Movement of Solutes between the Sieve Tubes and Surrounding Tissues in Willow. II. Pathways of Ion Transport from the Xylem to the Phloem', *J. Exp. Bot.*, **16**, 742–758

Honert, T. H. van den (1932). 'On the Mechanism of the Transport of Organic Materials in Plants', *Proc. K. Akad. Wet. (Amsterdam)*, **35**, 1104–1111

Hope, A. B. and Stevens, P. G. (1952). 'Electrical Potential Differences in Bean Roots and their Relation to Salt Uptake', *Aust. J. Sci. Res.*, **B.5**, 335–343

Horwitz, L. (1958). 'Some Simplified Mathematical Treatments of Translocation in Plants', *Plant Physiol. (Lancaster)*, **33**, 81–93

Huber, B., Schmidt, E. and Jahnel, H. (1937). 'Untersuchungen über den Assimilastrom. I. Mitteilung aus der Sachsischen Forstlichen Versuchsansalt

Tharandt, Abteilung für Botanik', *Tharandt. Forstl. Jb.*, **88**, 1017–1050

Hull, H. M. (1952). 'Carbohydrate Translocation in Tomato and Sugar Beet with Particular Reference to Temperature Effect', *Am. J. Bot.*, **39**, 661–669

Hylmö, B. (1953). 'Transpiration and Ion Absorption', *Physiologia Plant.*, **6**, 333–405

Hylmö, B. (1958). 'Passive Components in the Ion Absorption of the Plant. II. The Zonal Water Flow, Ion Passage and Pore Size in Roots of *Vicia faba*', *Physiologia Plant.*, **11**, 382–400

Ilan, I. (1962). 'A Specific Stimulatory Action of Indolyl-3-acetic Acid on Potassium Uptake by Plant Cells with a Concomitant Inhibition of Ammonium Uptake', *Nature (Lond.)*, **194**, 203–204

Jackson, J. E. and Weatherley, P. E. (1962). 'The Effect of Hydrostatic Pressure Gradients on the Movement of Potassium across the Root Cortex', *J. Exp. Bot.*, **13**, 128–143

Jacobs, W. P. (1961). 'The Polar Movement of Auxin in the Shoots of Higher Plants: Its Occurrence and Physiological Significance', in *Plant Growth Regulation* (Ed. R. M. Klein). Iowa State University Press, Ames, Iowa

Jarosch, R. (1964). 'Screw-Mechanical Basis of Protoplasmic Movement in Primitive Motile Systems', in *Cell Biology* (Ed. R. D. Allen and N. Kamiya). Academic Press, New York

Jarvis, P. and Thaine, R. (1971). 'Strands in Sections of Sieve Elements Cut in a Cryostat', *Nature New Biol. (Lond.)*, **232**, 236–237

Johnson, R. P. C. (1968). 'Microfilaments in Pores between Frozen-etched Sieve Elements', *Planta (Berl.)*, **81**, 314–332

Jones, H. and Eagles, J. E. (1962). 'Translocation of ^{14}C within and between Leaves', *Ann. Bot., N.S.*, **26**, 505–510

Jones, H., Martin, R. V. and Porter, H. K. (1959). 'Translocation of ^{14}C in Tobacco Following Assimilation of ^{14}CO$_2$ by a Single Leaf'; *Ann. Bot., N.S.*, **23**, 493–508

Jones, R. L. and Phillips, I. D. J. (1966). 'Organs of Gibberellin Synthesis in Light-grown Sunflower Plants', *Plant Physiol. (Lancaster)*, **41**, 1381–1386

Joy, K. W. (1964). 'Translocation in Sugar Beet. I. Assimilation of ^{14}CO$_2$ and Distribution of Materials from Leaves', *J. Exp. Bot.*, **15**, 485–494

Kamitsubo, E. (1966). 'Motile Protoplasmic Filaments in Cells of Characeae. II. Linear Fibrillar Structure and its Bearing on Protoplasmic Streaming', *Proc. Japan Acad.*, **42**, 640–643

Kamiya, N. (1953). 'The Motive Force Responsible for Protoplasmic Streaming in the Myxomycete Plasmodium', *A. Rep. Sci. Wks. Osaka Univ.*, **1**, 53–83

Kamiya, N. (1959). 'Protoplasmic Streaming', *Protoplasmatologia*, **8**, 1–199

Kamiya, N. (1960). 'Physics and Chemistry of Protoplasmic Streaming', *A. Rev. Plant Physiol.*, **11**, 323–340

Kamiya, N. (1968). 'The Mechanism of Cytoplasmic Movement in a Myxomycete Plasmodium', *Symp. Soc. Exp. Biol.*, **22**, 199–214

Kating, H. and Eschrich, W. (1964). 'Uptake, Incorporation and Transport of ^{14}C in *Cucurbita ficifolia*. II. Application of Bicarbonate ^{14}C to the Roots', *Planta (Berl.)*, **60**, 598–611

Kaufmann, M. R. and Kramer, P. J. (1967). 'Phloem Water Relations and Translocation', *Plant Physiol. (Lancaster)*, **42**, 191–194

Kendall, W. A. (1955). 'Effect of Certain Metabolic Inhibitors on Translocation of ^{32}P in Bean Plants', *Plant Physiol. (Lancaster)*, **30**, 347–350

Kende, H. (1965). 'Kinetin-like Factors in the Root Exudate of Sunflowers', *Proc. Natn. Acad. Sci. U.S.A.*, **53**, 130–137

Kennecke, M., Ziegler, H. and Fekete, M. A. R. (1971). 'Enzymaktivitäten im Siebröhrensaft von *Robinia pseudoacacia* L. und anderer Baumarten', *Planta*

(*Berl.*), **98**, 330–356

Kennedy, J. S. and Mittler, T. E. (1953). 'A Method for Obtaining Phloem Sap via the Mouth-parts of Aphids', *Nature (Lond.*), **171**, 528

Khan, A. A. and Sagar, G. R. (1966). 'Distribution of ¹⁴C-labelled Products of Photosynthesis during the Commercial Life of the Tomato Crop', *Ann. Bot. N.S.*, **30**, 727–743

King, E. N. (1971). 'The Movement of Solutes in the Bark and Wood of Willow', M.Sc. Thesis, University of Hull

Kleinig, H., Dörr, I., Weber, C. and Kollmann, R. (1971). 'Filamentous Proteins from Plant Sieve Tubes', *Nature New Biol. (Lond.*), **229**, 152–153

Kluge, M., Becker, D. and Ziegler, H. (1970). 'Untersuchungen über ATP und andere Organische Phosphorverbidungen im Siebröhrensaft von *Yucca flaccida* und *Salix triandra*,' *Planta (Berl.*), **91**, 68–79

Kluge, M., Reinhard, E. and Ziegler, H. (1964). 'Gibberellinaktivität von Siebrohrensäften', *Naturwissenschaften*, **6**, 145–146

Kluge, M. and Ziegler, H. (1964). 'Der ATP-Gehlat der Siebröhrensäfte von Laubbäumen', *Planta (Berl.*), **61**, 167–177

Kollmann, R. (1960). 'Untersuchungen über das Protoplasma der Siebröhren von *Passiflora caereua*. II. Electronenopische Untersuchungen', *Planta (Berl.*), **55**, 67–107

Kollmann, R. (1964). 'On the Fine Structure of the Sieve Element Protoplast', *Phytomorphology*, **14**, 247–264

Kollmann, R. (1965). 'Zur Lokalisierung der Funktionstüchtigen Siebzellen im Sekundären Phloem von *Metasequoia glyptostroboides*', *Planta (Berl.*), **65**, 173–179

Kollmann, R. (1967). 'Autoradiographischer Nachweis der Assimilat-Transportbahn im Sekundären Phloem von *Metasequoia glyptostroboides*', *Z. Pflanzenphysiol.*, **56**, 401–409

Kollmann, R. and Dörr, I. (1966). 'Lokalisierung Funktionstüchtiger Siebzellen bei *Juniperis communis* mit Hilfe von Aphiden', *Z. Pflanzenphysiol.*, **55**, 131–141

Kollmann, R., Dörr, I. and Kleinig, H. (1970). 'Protein Filaments—Structural Components of the Phloem Exudate', *Planta (Berl.*), **95**, 86–94

Kollmann, R. and Schumacher, W. (1964). 'Uber die Feinstrukter des Phloems von *Metasequoia glyptostroboides* und seine Jahreszeitlichen Veranderungen. V. Die Differenzierung der Siebzellen in Verhaufe einer Vegetationsperiode', *Planta (Berl.*), **63**, 155–190

Kramer, P. J. (1949). *Plant and Soil Water Relationships*. McGraw-Hill, New York

Kursanov, A. L. (1963). 'Metabolism and the Transport of Organic Substances in the Phloem', *Adv. Bot. Res.*, **1**, 209–274

Laties, G. C. and Budd, K. (1964). 'The Development of Differential Permeability in Isolated Steles of Corn Roots', *Proc. Natn. Acad. Sci. U.S.A.*, **52**, 462–469

Lauchli, A. (1967). 'Untersuchungen über Verteiling und Transport von Ionen in Pflanzengeweben mit der Rontgenmikrosonde. I. Versuche an Vegitativen Organen von *Zea mays*', *Planta (Berl.*), **75**, 185–206

Lawton, J. R. S. and Canny, M. J. (1970). 'The Proportion of Sieve Elements in the Phloem of Some Tropical Trees', *Planta (Berl.*), **95**, 351–354

Leach, R. W. A. and Wareing, P. F. (1967). 'Distribution of Auxin in Horizontal Woody Stems in Relation to Gravimorphism', *Nature (Lond.*), **214**, 1025–1027

Lee, D. R. (1972). 'The Possible Significance of Filaments in Sieve Elements', *Nature (Lond.*), **235**, 286

Lee, D. R., Arnold, D. C. and Fensom, D. S. (1971). 'Some Microscopical Observations of Functioning Sieve Tubes of *Heracleum* using Nomarski Optics',

J. Exp. Bot., **22**, 25–38

Lenton, J. R. Bowen, M. R. and Saunders, P. F. (1968). 'Detection of Abscisic Acid in the Xylem Sap of Willow (*Salix viminalis* L.) by Gas-Liquid Chromatography', *Nature (Lond.)*, **220**, 86–87

Leopold, A. C. and Lam, S. L. (1961). 'Polar Transport of Three Auxins', in *Plant Growth Regulation* (Ed. R. M. Klein). Iowa State University Press, Ames, Iowa

Lepp, N. W. and Peel, A. J. (1970). 'Some Effects of IAA and Kinetin on the Movement of Sugars in the Phloem of Willow', *Planta (Berl.)*, **90**, 230–235

Lepp, N. W. and Peel, A. J. (1971a). 'Patterns of Translocation and Metabolism of ¹⁴C-labelled IAA in the Phloem of Willow', *Planta (Berl.)*, **96**, 62–73

Lepp, N. W. and Peel, A. J. (1971b). 'Influence of IAA upon the Longitudinal and Tangential Movement of Labelled Sugars in the Phloem of Willow', *Planta (Berl.)*, **97**, 50–61

Lepp, N. W. and Peel, A. J. (1971c). 'Distribution of Growth Regulators and Sugars by the Tangential and Radial Transport Systems of Stem Segments of Willow', *Planta (Berl.)*, **99**, 275–282

Lester, H. H. and Evert, R. F. (1965). 'Acid Phosphatase Activity in Sieve Tube Members of *Tilia americana*,' *Planta (Berl.)*, **65**, 180–185

Little, C. H. A. and Blackman, G. E. (1963). 'The Movement of Growth Regulators in Plants. III. Comparative Studies of the Transport in *Phaseolus vulgaris*', *New Phytol.*, **62**, 173–197

Luttge, U. and Laties, G. C. (1967). 'Absorption and Transport by Isolated Steles of Maize Roots in Relation to the Dual Mechanisms of Ion Absorption', *Planta (Berl.)*, **74**, 173–187

Luttge, U. and Weigl, J. (1962). 'Mikroautoradiographische Untersuchungen der Aufnahme und des Transportes von ³⁵SO₄ und ⁴⁵Ca in Keimwurzeln von *Zea mays* L. und *Pisum sativum* L.', *Planta (Berl.)*, **58**, 113–126

McComb, A. J. (1964). 'The Stability and Movement of Gibberellic Acid in Pea Seedlings', *Ann. Bot., N.S.*, **28**, 669–687

McCready, C. C. (1963). 'Movement of Growth Regulators in Plants. I. Polar Transport of 2,4-Dichlorophenoxyacetic Acid in Segments from the Petioles of *Phaseolus vulgaris*,' *New Phytol*, **62**, 3–18

McCready, C. C. (1966). 'Translocation of Growth Regulators', *A. Rev. Plant Physiol.*, **17**, 283–294

McCready, C. C. and Jacobs, W. P. (1963). 'Movement of Growth Regulators in Plants. II. Polar Transport of Radioactivity from Indoleacetic Acid-(¹⁴C) and 2,4-Dichlorophenoxyacetic Acid-(¹⁴C) in Petioles of *Phaseolus vulgaris*', *New Phytol.*, **62**, 19–34

McNairn, R. B. and Currier, H. B. (1968). 'Translocation Blockage by Sieve Plate Callose', *Planta (Berl.)*, **82**, 369–380

MacRobbie, E. A. C. (1971). 'Phloem Translocation. Facts and Mechanisms: a Comparative Survey', *Biol. Rev.*, **46**, 429–481

Mason, T. G. and Lewin, C. J. (1926). 'On the Rate of Carbohydrate Transport in the Greater Yam *Dioscorea alata*', *Sci. Proc. R. Dublin Soc.*, **18**, 203–205

Mason, T. G. and Maskell, E. J. (1928a). 'Studies on the Transport of Carbohydrates in the Cotton Plant. I. A Study of Diurnal Variation in the Carbohydrates of Leaf Bark and Wood, and of the Effects of "Ringing" ', *Ann. Bot.*, **42**, 189–253

Mason, T. G. and Maskell, E. J. (1928b). 'Studies on the Transport of Carbohydrates in the Cotton Plant. II. The Factors Determining the Rate and Direction of Movement of Sugars', *Ann. Bot.*, **42**, 571–636

Mason, T. G. and Maskell, E. J. (1929). 'Studies on the Transport of Nitrogenous Substances in the Cotton Plant. I. Preliminary Observations on the

Downward Transport of Nitrogen in the Stem', *Ann. Bot.*, **43**, 205–232

Mason, T. G. and Maskell, E. J. (1931). 'Further Studies on Transport in the Cotton Plant. I. Preliminary Observations on the Transport of Phosphorus, Potassium and Calcium', *Ann. Bot.*, **45**, 125–174

Mason, T. G., Maskell, E. J. and Phillis, E. (1936). 'Further Studies on Transport in the Cotton Plant. III. Concerning the Independence of Solute Movement in the Phloem', *Ann. Bot.*, **50**, 23–58

Mason, T. G. and Phillis, E. (1936). 'Further Studies on Transport in the Cotton Plant. V. Oxygen Supply and the Activation of Diffusion', *Ann. Bot.*, **50**, 455–499

Mehta, A. S. and Spanner, D. C. (1962). 'The Fine Structure of the Sieve Tubes of the Petiole of *Nymphoides peltatum* (Gmel) O. Kunze', *Ann. Bot.*, *N.S.*, **26**, 291–299

Milburn, J. A. (1970). 'Phloem Exudation from Caster Bean: Induction by Massage', *Planta (Berl.)*, **95**, 272–276

Milburn, J. A. (1971). 'An Anlaysis of the Response in Phloem Exudation on Application of Massage to *Ricinus*', *Planta (Berl.)*, **100**, 143–154

Mishra, U. and Spanner, D. C. (1970). 'The Fine Structure of the Sieve Tubes of *Salix caprea* (L.) and its relation to the Electro-osmotic Theory', *Planta (Berl.)*, **90**, 43–56

Mitchell, J. W. and Brown, J. W. (1945). 'Movement of 2,4-Dichlorophenoxyacetic Acid Stimulus and its Relation to the Translocation of Organic Food Materials in Plants', *Bot. Gaz.*, **107**, 393–407

Mitchell, J. W. and Martin, W. E. (1937). 'Effect of Indoleacetic Acid on Growth and Composition of Etiolated Bean Plants', *Bot. Gaz.*, **99**, 171–183

Mittler, T. E. (1954). 'Studies on the Feeding and Nutrition of *Tuberolachnus salignus* (Gmelin)', Ph.D. Thesis, University of Cambridge

Mittler, T. E. (1958). 'Studies on the Feeding and Nutrition of *Tuberolachnus salignus* (Gmelin). II. The Nitrogen and Sugar Composition of Ingested Phloem Sap and Excreted Honeydew', *J. Exp. Biol.*, **35**, 74–84

Moorby, J., Ebert, M. and Evans, N. T. S. (1963). 'The Translocation of ^{14}C-labelled Assimilate in the Soybean', *J. Exp. Bot.*, **14**, 210–220

Moose, C. A. (1938). 'Chemical and Spectroscopic Analysis of Plants', *Plant Physiol. (Lancaster)*, **13**, 365–380

Morris, D. A., Briant, R. E. and Thomson, P. G. (1969). 'The Transport and Metabolism of ^{14}C-labelled Indoleacetic Acid in Intact Pea Seedlings', *Planta (Berl.)*, **89**, 178–197

Mortimer, D. C. (1961). 'Translocation of Photosynthetic Products in Sugar Beet Petioles', *Plant Physiol. (Lancaster)*, **36**, Proceedings xxxiv

Mothes, K. and Engelbrecht, L. (1961). 'Kinetin-induced Directed Transport of Substances in Excised Leaves in the Dark', *Phytochemistry* **1**, 58–62

Mothes, K., Engelbrecht, L. and Schütte, H. R. (1961). 'Über die Akkumulation von α-aminobuttersäure in Blattgewebe unter dem Einfluss von Kinetin', *Physiologia Plant.*, **14**, 72–75

Mullins, M. G. (1970). 'Hormone-directed Transport of Assimilates in Decapitated Internodes of *Phaseolus vulgaris* L.', *Ann. Bot.*, *N.S.*, **34**, 897–909

Münch, E. (1930). *Die Stoffbewegungen in der Pflanze*. Fischer, Jena

Münch, E. (1937). 'Versuche über Wege und Richtungen der Stoffbewegungen im Baum', *Forstwiss. Zbl.*, **59**, 305–324

Murmanis, L. and Evert, R. F. (1966). 'Some Aspects of Sieve Cell Ultrastructure in *Pinus strobus*', *Am. J. Bot.*, **53**, 1065–1078

Nagai, R. and Rebhun, L. I. (1966). 'Cytoplasmic Microfilaments in Streaming *Nitella* Cells', *J. Ultrastruct. Res.*, **14**, 571–589

Naqui, S. M. and Gordon, S. A. (1965). 'Auxin Transport in Flowering and Vegetative Shoots of *Coleus blumei* Benth.', *Plant Physiol. (Lancaster)*, **40**, 116–118

Nelson, C. D. (1962). 'The Translocation of Organic Compounds in Plants', *Can. J. Bot.*, **40**, 757–770

Nelson, C. D. (1963). 'Effect of Climate on the Distribution and Translocation of Assimilates', in *Environmental Control of Plant Growth* (Ed. L. T. Evans). Academic Press, New York

Nelson, C. D. and Gorham, P. R. (1937). 'Translocation of Radioactive Sugars in the Stems of Soyabean Seedlings', *Can. J. Bot.*, **35**, 703–713

Nelson, C. D., Perkins, H. J. and Gorham, P. R. (1958). 'Note on a Rapid Translocation of Photosynthetically Assimilated ^{14}C out of the Primary Leaf of the Young Soybean Plant', *Can. J. Biochem. Biophys.*, **36**, 1277–1279

Niedergang-Kamien, E. and Leopold, A. C. (1957). 'Inhibitors of Polar Auxin Transport', *Physiologia Plant.*, **10**, 29–38

Nitsch, J. P. and Nitsch, C. (1965). 'Présence du Phytokinins et autres Substances de Croissance dans la sève d'*Acer saccharum* et de *Vitis vinifera*', *Bull. Soc. Bot. France*, **112**, 11–18

Northcote, D. H. and Wooding, F. P. B. (1966). 'Development of Sieve Tubes in *Acer pseudoplatanus*', *Proc. R. Soc. (B)*, **163**, 524–536

Osborne, D. J. and Hallaway, M. (1961). 'The Role of Auxins in the Control of Leaf Senescence. Some Effects of Local Applications of 2,4-D on Carbon and Nitrogen Metabolism', in *Plant Growth Regulation* (Ed. R. M. Klein). Iowa State University Press, Ames, Iowa

Oserkowsky, J. (1942). 'Polar and Apolar Transport of Auxin in Woody Stems', *Am. J. Bot.*, **29**, 858–866

Overbeek, J. van (1942). 'Water Uptake by Excised Root Systems of the Tomato Due to Non-Osmotic Forces', *Am. J. Bot.*, **29**, 677–683

Palmquist, E. M. (1938). 'The Simultaneous Movement of Carbohydrates and Fluorescein in Opposite Directions in the Phloem', *Am. J. Bot.*, **25**, 97–105

Parker, J. (1964). 'Sieve Tube Strands in Tree Bark', *Nature (Lond.)*, **202**, 926–927

Parthasarathy, M. V. and Tomlinson, P. B. (1967). 'Anatomical Features of Metaphloem in Stems of *Sabal, Cocos* and two other Palms', *Am. J. Bot.*, **54**, 1143–1151

Peel, A. J. (1959). 'Studies on the Physiology of the Sieve Tube in Higher Plants', Ph.D. Thesis, University of Nottingham

Peel, A. J. (1963). 'The Movement of Ions from the Xylem Solution into the Sieve Tubes of Willow', *J. Exp. Bot.*, **14**, 438–447

Peel, A. J. (1964). 'Tangential Movement of ^{14}C-labelled Assimilates in Stems of Willow', *J Exp. Bot.*, **15**, 104–113

Peel, A. J. (1965a). 'On the Conductivity of the Xylem in Trees', *Ann. Bot., N.S.*, **29**, 119–130

Peel, A. J. (1965b). 'The Effect of Changes in the Diffusion Potential of the Xylem Water on Sieve-tube Exudation from Isolated Stem Segments', *J. Exp. Bot.*, **16**, 249–260

Peel, A. J. (1966). 'The Sugars Concerned in the Tangential Movement of ^{14}C-labelled Assimilates in Willow', *J. Exp. Bot.*, **17**, 156–164

Peel, A. J. (1967). 'Demonstration of Solute Movement from the Extracambial Tissues into the Xylem Stream in Willow', *J. Exp. Bot.*, **18**, 600–606

Peel, A. J. (1970). 'Further Evidence for the Relative Immobility of Water in Sieve Tubes of Willow', *Physiologia Plant.*, **23**, 667–672

Peel, A. J. (1972a). 'The Control of the Rate and Direction of Phloem Transport', in *Crop Processes in Controlled Environments* (Ed. A. R. Rees, K. E. Cockshull,

D. W. Hand and R. G. Hurd). Academic Press, London

Peel, A. J. (1972b). 'The Control of Solute Movement into Sieve Elements', *Pestic. Sci.*, **3**, 631–641

Peel, A. J., Field, R. J., Coulson, C. L. and Gardner, D. C. J. (1969). 'Movement of Water and Solutes in Sieve Tubes of Willow in Response to Puncture by Aphid Stylets. Evidence against a Mass Flow of Solution', *Physiologia Plant.*, **22**, 768–775

Peel, A. J. and Ford, J. (1968). 'The Movement of Sugars into the Sieve Elements of Bark Strips of Willow. II. Evidence for Two Pathways from the Bathing Solution', *J. Exp. Bot.*, **19**, 370–380

Peel, A. J and Ho, L C. (1970). 'Colony Size of *Tuberolachnus salignus* Gmelin in Relation to Mass Transport of ^{14}C-labelled Assimilates from the Leaves in Willow', *Physiologia Plant.*, **23**, 1033–1038

Peel, A. J. and Weatherley, P. E. (1959). 'Composition of Sieve Tube Sap', *Nature (Lond.)*, **184**, 1955–1956

Peel, A. J. and Weatherley, P. E. (1962). 'Studies in Sieve Tube Exudation through Aphid Mouthparts. I. The Effects of Light and Girdling', *Ann. Bot.*, *N.S.*, **26**, 633–646

Peel, A. J. and Weatherley, P. E. (1963). 'Studies in Sieve Tube Exudation through Aphid Mouthparts. II. The Effects of Pressure Gradients in the Wood and Metabolic Inhibitors', *Ann. Bot.*, *N.S.*, **27**, 197–211

Phillips, I. D. J. and Jones, R. L. (1964). 'Gibberellin-like Activity in Bleeding Sap of Root Systems of *Helianthus annuus* Detected by a New Dwarf Pea Epicotyl Assay and Other Methods', *Planta (Berl.)*, **63**, 269–278

Phillis, E. and Mason, T. G. (1933). 'Studies on the Transport of Carbohydrates in the Cotton Plant. III. The Polar Distribution of Sugar in the Foliage Leaf', *Ann. Bot.*, **47**, 585–634

Phillis, E. and Mason, T. G. (1936). 'Further Studies on Transport in the Cotton Plant. IV. On the Simultaneous Movement of Solutes in Opposite Directions through the Phloem', *Ann. Bot.*, **50**, 161–174

Pilet, P. E. (1965). 'Polar Transport of Radioactivity from ^{14}C-labelled Indolylacetic Acid in Stems of *Lens culinaris*', *Physiologia Plant.*, **18**, 687–702

Plaut, Z. and Reinhold, L. (1965). 'The Effect of Water Stress on ^{14}C-sucrose Transport in Bean Plants', *Aust. J. Biol. Sci.*, **18**, 1143–1155

Pollard, T. D., Shelton, E., Weihung, R. R. and Korn, E. D. (1970). 'Ultrastructural Characterisation of F-Actin Isolated from *Acanthamoeba castellanii* and Identification of Cytoplasmic Filaments as F-Actin by Reaction with Rabbit Heavy Meromyosin', *J. Molec. Biol.*, **50**, 91–97

Preston, R. D. (1952). 'Movement of Water in Higher Plants', in *Deformation and Flow in Biological Systems* (Ed. A. Frey-Wyssling). North-Holland, Amsterdam

Prokofyev, A. A., Zhadanova, L. P. and Sobolev, A. H. (1957). 'Certain Regularities in the Flow of Substances from Leaves into Reproductive Organs', *Fiziologiya Rast.*, **4**, 425–431

Qureshi, F. A. and Spanner, D. C. (1973). 'The Effect of Nitrogen on the Movement of Tracers down the Stolon of *Saxifraga farmentosa*, with some observations on the influence of Light', *Planta (Berl.)*, **110**, 131–144

Qureshi, F. A. and Spanner, D. C. (1973). 'The Influence of Dinitrophenol on Phloem Transport along the Stolon of *Saxifraga farmentosa*', *Planta (Berl.)*, **111**, 1–12

Qureshi, F. A. and Spanner, D. C. (1973). 'Cyanide Inhibition of Phloem Transport along the Stolon of *Saxifraga farmentosa* L.', *J. Exp. Bot.*, **24**, 751–762

Rabideau, G. S. and Burr, G. O. (1945). 'The Use of ^{13}C Isotope as a Tracer for Transport Studies in Plants', *Am. J. Bot.*, **32**, 349–356

Roberts, B. R. (1964). 'Effect of Water Stress on the Translocation of Photo-synthetically Assimilated Carbon-14 in Yellow Poplar', in *The Formation of Wood in Forest Trees* (Ed. M. H. Zimmermann). Academic Press, New York

Rohrbaugh, L. M. and Rice, E. L. (1949). 'Effect of Application of Sugar on the Translocation of Sodium 2,4-Dichlorophenoxyacetate by Bean Plants in the Dark', *Bot. Gaz.*, **110**, 85–89

Rouschal, E. (1941). 'Untersuchungen über die Protoplasmatik und Funktion der Siebröhren', *Flora (Jena)*, **35**, 135–200

Russell, R. S. and Barber, D. A. (1960). 'The Relationship between Salt Uptake and the Absorption of Water by Intact Plants', *A. Rev. Plant Physiol.*, **11**, 127–140

Russell, R. S. and Shorrocks, V. M. (1959). 'The Relationship between Trans-piration and Absorption of Inorganic Ions by Intact Plants', *J. Exp. Bot.*, **10**, 301–316

Sabnis, D. D., Hirshberg, G. and Jacobs, W. P. (1969). 'Radioautographic Analysis of the Distribution of Label from ^{3}H-Indoleacetic Acid Supplied to Isolated *Coleus* Internodes', *Plant Physiol. (Lancaster)*, **44**, 27–36

Sacher, J. A., Hatch, M. D. and Glasziou, K. T. (1963). 'Sugar Accumulation Cycle in Sugar Cane. III. Physical and Metabolic Aspects of Cycle in Immature Storage Tissues', *Plant Physiol. (Lancaster)*, **38**, 348–354

Sachs, T. and Thimann, K. V. (1964). 'Release of Lateral Buds from Apical Dominance', *Nature (Lond.)*, **201**, 939–940

Scholander, P. F., Bradstreet, E. D., Hammel, H. T. and Hemmingsen, E. A. (1966). 'Sap Concentrations in Halophytes and Some Other Plants', *Plant Physiol. (Lancaster)*, **41**, 529–532

Scholander, P. F., Love, W. E. and Kanwisher, J. W. (1955). 'The Rise of Sap in Tall Grapevines', *Plant Physiol. (Lancaster)*, **30**, 93–104

Scholander, P. F., Ruud, B. and Levestad, H. (1957). 'The Rise of Sap in a Tropical Liana', *Plant Physiol. (Lancaster)*, **32**, 1–6

Schumacher, W. (1933). 'Untersuchungen über die Wanderung des Fluoreszeins in die Siebröhren', *Jb. Wiss. Bot.*, **77**, 685–732

Schumacher, W. (1939). 'Über die Plasmolysierbarkeit der Sierbröhren', *Jb. Wiss. Bot.*, **88**, 545–553

Seth, A. K. and Wareing, P. F. (1967). 'Hormone-directed Transport of Meta-bolites and its Possible Role in Plant Senescence', *J. Exp. Bot.*, **18**, 65–77

Shih, C. Y. and Currier, H. B. (1969). 'Fine Structure of Phloem Cells in Relation to Translocation in the Cotton Seedling', *Am. J. Bot.*, **56**, 464–472

Shiroya, M., Lister, G. R., Nelson, C. D. and Krotkov, G. (1961). 'Translocation of ^{14}C in Tobacco at Different Stages of Development Following Assimilation of $^{14}CO_2$ by a Single Leaf', *Can. J. Bot.*, **39**, 855–886

Siddiqui, A. W. and Spanner, D. C. (1970). 'The State of the Pores in Function-ing Sieve Plates', *Planta (Berl.)*, **99**, 181–189

Singh, A. P. and Srivastava, L. M. (1972). 'The Fine Structure of Corn Phloem', *Can. J. Bot.*, **50**, 839–846

Slatyer, R. O. (1967). *Plant–Water Relationships*. Academic Press, London

Spanner, D. C. (1958). 'The Translocation of Sugar in Sieve Tubes', *J. Exp. Bot.*, **9**, 332–342

Spanner, D. C. (1962). 'A Note on the Velocity and Energy Requirement of Translocation', *Ann. Bot., N.S.*, **26**, 511–516

Spanner, D. C. (1970). 'The Electro-osmotic Theory of Phloem Transport in the Light of Recent Measurements on *Heracleum* Phloem', *J. Exp. Bot.*, **21**, 325–334

Spanner, D. C. and Jones, R. L. (1970). 'The Sieve Tube Wall and its Relation to Translocation', *Planta (Berl.)*, **92**, 64–72

Srivastava, L. M. (1963). 'Secondary Phloem in the Pinaceae', *Univ. Calif. Publs. Bot.*, **34**, 1–142

Stout, P. R. (1945). 'Translocation of the Reproductive Stimulus in Sugar Beets', *Bot. Gaz.*, **107**, 86–95

Stout, P. R. and Hoagland, D. R. (1939). 'Upward and Lateral Movement of Salt in Certain Plants as Indicated by Radioactive Isotopes of Potassium, Sodium and Phosphorus Absorbed by the Roots', *Am. J. Bot.*, **26**, 320–324

Sunderam, A. (1965). 'A Preliminary Investigation of the Penetration and Translocation of 2,4,5-T in Some Tropical Trees', *Weed Res.*, **5**, 213–225

Swanson, C. A. and Böhning, R. H. (1951). 'The Effect of Petiole Temperature on the Translocation of Carbohydrates from Bean Leaves', *Plant Physiol. (Lancaster)*, **26**, 557–564

Swanson, C. A. and El-Shishiny, E. D. H. (1958). 'Translocation of Sugars in the Concord Grape', *Plant Physiol. (Lancaster)*, **33**, 33–37

Swanson, C. A. and Geiger, D. R. (1967). 'Time Course of Low Temperature Inhibition of Sucrose Translocation in Sugar Beets', *Plant Physiol. (Lancaster)*, **42**, 751–756

Swanson, C. A. and Whitney, J. B. (1953). 'Studies on the Translocation of Foliar Applied ^{32}P and Other Radioisotopes in Bean Plants', *Am. J. Bot.*, **40**, 816–823

Tammes, P. M. L. and Die, J. van (1964). 'Studies on Phloem Exudation from *Yucca flaccida* Haw. I. Some Observations on the Phenomenon of Bleeding and the Composition of the Exudate', *Acta. Bot. Neerl.*, **13**, 76–83

Tammes, P. M. L. and Die, J. van (1966). 'Studies on Phloem Exudation from *Yucca flaccida* Haw. IV. Translocation of Macro- and Micro-nutrients by the Phloem Sap Stream', *Proc. K. Ned. Akad. Wet. (Amsterdam)*, **69**, 656–659

Tammes, P. M. L., Die, J. van and Ie, T. S. (1971). 'Studies on Phloem Exudation from *Yucca flaccida* Haw. VIII. Fluid Mechanics and Exudation', *Acta. Bot. Neerl.*, **20**, 245–252

Tammes, P. M. L. and Ie, T. S. (1971). 'Studies on Phloem Exudation from *Yucca flaccida* Haw. IX. Passage of Carbon Black Particles through Sieve Plate Pores of *Yucca flaccida* Haw', *Acta. Bot. Neerl.*, **20**, 309–317

Tammes, P. M. L., Vonk C. R. and Die, J. van (1969). 'Studies on Phloem Exudation from *Yucca flaccida* Haw. VII. The Effect of Cooling on Exudation', *Acta. Bot. Neerl.*, **18**, 224–246

Tamulevich, S. R. and Evert, R. F. (1966). 'Aspects of Sieve Element Ultrastructure in *Primula obconica*', *Planta (Berl.)*, **69**, 319–337

Thaine, R. (1961). 'Transcellular Strands and Particle Movement in Mature Sieve Tubes', *Nature (Lond.)*, **192**, 772–773

Thaine, R. (1962). 'A Translocation Hypothesis Based on the Structure of Plant Cytoplasm', *J. Exp. Bot.*, **13**, 152–160

Thaine, R. (1964). 'The Protoplasmic Streaming Theory of Phloem Transport', *J. Exp. Bot.*, **15**, 470–484

Thaine, R. (1969). 'Movement of Sugars through Plants by Cytoplasmic Pumping', *Nature (Lond.)*, **222**, 873–875

Thaine, R., Ovenden, S. L. and Turner, J. S. (1959). 'Translocation of Labelled Assimilates in the Soybean', *Aust. J. Biol Sci.*, **12**, 349–371

Thaine, R., Probine, M. C. and Dyer, P. Y. (1967). 'The Existence of Transcellular Strands in Sieve Elements', *J. Exp. Bot.*, **18**, 110–127

Thimann, K. V. and Sweeney, B. M. (1937). 'The Effect of Auxins upon Protoplasmic Streaming', *J. Gen. Physiol.*, **21**, 123–135

Thrower, S. L. (1962). 'Translocation of Labelled Assimilates in Soybean. II. The Pattern of Translocation in Intact and Defoliated Plants', *Aust. J. Biol.*

Sci., **15**, 629–649

Thrower, S. L. (1965). 'Translocation of Labelled Assimilates in the Soybean. IV. Some Effects of Low Temperature on Translocation', *Aust. J. Biol. Sci.*, **18**, 449–461

Ting, T. (1963). 'Translocation of Assimilates of Leaf of Main Stem in Relation to Phyllotaxis in the Cotton Plant', *Acta. Biol. Exp. Sin.*, **8**, 656–663

Tingley, M. A. (1944). 'Concentration Gradients in Plant Exudates with Reference to the Mechanism of Translocation', *Am. J. Bot.*, **31**, 30–38

Tomlinson, P. B. (1964). 'Stem Structure in Arborescent Monocotyledons', in *The Formation of Wood in Forest Trees* (Ed. M. H. Zimmermann). Academic Press, New York

Trip, P. and Gorham, P. R. (1968a). 'Bidirectional Translocation of Sugars in Sieve Tubes of Squash Plants', *Plant Physiol. (Lancaster)*, **43**, 877–882

Trip, P. and Gorham, P. R. (1968b). 'Translocation of Sugar and Tritiated Water in Squash Plants', *Plant Physiol. (Lancaster)*, **43**, 1845–1849

Trip, P., Nelson, C. D. and Krotkov, G. (1965). 'Selective and Preferential Translocation of ^{14}C-labelled Sugars in White Ash and Lilac', *Plant Physiol. (Lancaster)*, **40**, 740–747

Turner, E. R. (1960). 'The Movement of Organic Nitrogen Compounds in Plants', *Ann. Bot., N.S.*, **24**, 387–396

Turner, T. S., Macrae, J. and Grant-Lipp, P. (1954). 'Auxin and Protoplasmic Streaming—an Oxygen Effect', *Bot. Congr. Paris, Section 11*, 152–155

Tyree, M. T. (1970). 'The Symplast Concept. A General Theory of Symplastic Transport According to the Thermodynamics of Irreversible Processes', *J. Theor. Biol.*, **26**, 181–214

Tyree, M. T. and Fensom, D. S. (1970). 'Some Experimental and Theoretical Observations Concerning Mass Flow in the Vascular Bundles of *Heracleum*', *J. Exp. Bot.*, **21**, 304–324

Vernon, L. P. and Aronoff, S. (1952). 'Metabolism of Soybean Leaves. IV. Translocation from Soybean Leaves', *Arch. Biochem. Biophys.*, **36**, 383–398

Walker, T. S. and Thaine, R. (1971). 'Proteins and Fine Structural Components in Exudate from Sieve Tubes in *Cucurbita pepo* Stems', *Ann. Bot., N.S.*, **35**, 773–790

Wangermann, E. (1970). 'Autoradiographic Localisation of Soluble and Insoluble ^{14}C from (^{14}C) Indolylacetic Acid Supplied to Isolated *Coleus* Internodes', *New Phytol.*, **69**, 919–927

Wanner, H. (1953). 'Enzyme der Glykolyse in Phlöemsaft', *Ber. Schweiz. Bot. Ges.*, **63**, 201–212

Wardlaw, I. F. (1967). 'The Effect of Water Stress on Translocation in Relation to Photosynthesis and Growth. I. Effect During Grain Development in Wheat', *Aust. J. Biol. Sci.*, **20**, 25–39

Wardlaw, I. F. (1968). 'The Control and Pattern of Movement of Carbohydrates in Plants', *Bot. Rev.*, **34**, 79–105

Wardlaw, I. F. (1969). 'The Effect of Water Stress on Translocation in Relation to Photosynthesis and Growth', *Aust. J. Biol. Sci.*, **22**, 1–16

Wardlaw, I. F. and Porter, H. K. (1967). 'The Redistribution of Stem Sugars in Wheat During Grain Development', *Aust. J. Biol. Sci.*, **20**, 309–318

Wareing, P. F. (1970). 'Growth and its Co-ordination in Trees', in *Physiology of Tree Crops* (Ed. L. C. Luckwill and C. V. Cutting). Academic Press, London

Wark, M. C. and Chambers, T. C. (1965). 'Fine Structure of the Phloem of *Pisum sativum*. I. The Sieve Element Ontogeny', *Aust. J. Bot.*, **13**, 171–183

Weatherley, P. E. (1963). 'The Pathway of Water Movement across the Root Cortex and Leaf Mesophyll of Transpiring Plants', in *The Water Relations of Plants* (Ed. A. J. Rutter and F. H. Whitehead). Blackwell, Oxford

Weatherley, P. E. and Johnson, R. P. C. (1968). 'The Form and Function of the Sieve Tube: A Problem in Reconciliation', *Int. Rev. Cytol.*, **24**, 149–192

Weatherley, P. E., Peel, A. J. and Hill, G. P. (1959). 'The Physiology of the Sieve Tube. Preliminary Investigations Using Aphid Mouth Parts', *J. Exp. Bot.*, **10**, 1–16

Weatherley, P. E. and Watson, B. T. (1969). 'Some Low-Temperature Effects on Sieve Tube Translocation in *Salix viminalis*', *Ann. Bot.*, *N.S.*, **33**, 845–853

Webb, J. A. (1967). 'Translocation of Sugars in *Cucurbita melopepo*. IV. Effects of Temperature Change', *Plant Physiol. (Lancaster)*, **42**, 881–885

Webb, J. A. and Gorham, P. R. (1965a). 'Radial Movement of ^{14}C-Translocates from Squash Phloem', *Can. J. Bot.*, **43**, 97–103

Webb, J. A. and Gorham, P. R. (1965b). 'The Effect of Node Temperature on Assimilation and Translocation of ^{14}C in the Squash', *Can. J. Bot.*, **43**, 1009–1020

Weij, H. G. van der (1932). 'Der Mechanismus des Wuchsstofftransportes. I', *Rec. Trav. Bot. Neerl.*, **29**, 380–496

Weij, H. G. van der (1934). 'Der Mechanismus des Wuchsstofftransportes. II', *Rec. Trav. Bot. Neerl.*, **31**, 810–857

Went, F. W. (1928). 'Wuchstoff und Wachstum', *Rec. Trav. Bot. Neerl.*, **25**, 1–116

Went, F. W. (1936). 'Allgemeine Betrachtungen über das Auxinproblem', *Biol. Zbl.*, **56**, 449–463

Went, F. W. and Hull, H. M. (1949). 'The Effect of Temperature on Translocation of Carbohydrate in the Tomato Plant', *Plant Physiol. (Lancaster)*, **24**, 505–526

Went, F. W. and White, R. (1939). 'Experiments on the Transport of Auxin', *Bot. Gaz.*, **100**, 465–484

Whitehouse, R. L. and Zalik, S. (1967). 'Translocation of Indole-3-acetic acid-1-^{14}C and Tryptophan-1-^{14}C in Seedlings of *Phaseolus coccineus* L. and *Zea mays* L.', *Plant Physiol. (Lancaster)*, **42**, 1363–1372

Whittle, C. M. (1964). 'Translocation and Temperature', *Ann. Bot.*, *N.S.*, **28**, 339–344

Wiebe, H. H. and Kramer, P. J. (1954). 'Translocation of Radioactive Isotopes from Various Regions of Roots of Barley Seedlings', *Plant Physiol. (Lancaster)*, **29**, 342–348

Wiersum, L. K., Vonk, C. A. and Tammes, P. M. L. (1971). 'Movement of ^{45}Ca in the Phloem of *Yucca*', *Naturwissenschaften*, **99**, 104–105

Willenbrink, J. (1957). 'Über die Hemmung des Stofftransports in den Siebröhren durch lokale Inacktivierung Verschiedener Atmungerzyme', *Planta (Berl.)*, **48**, 269–342

Wooding, F. B. P. (1966). 'The Development of the Sieve Elements of *Pinus pinea*', *Planta (Berl.)*, **69**, 230–243

Wooding, F. B. P. and Northcote, D. H. (1965). 'The Fine Structure and Development of the Companion Cell of the Phloem of *Acer pseudoplatanus*', *J. Cell. Biol.*, **24**, 117–128

Wray, F. J. and Richardson, J. A. (1964). 'Paths of Water Transport in Higher Plants', *Nature (Lond.)*, **202**, 415–416

Yang, T. T. (1961). 'A Study of Translocation of Photosynthate in Sugar Cane Plants with ^{14}C and ^{32}P ', *Rep. Taiwan Sugar Exp. Stn.*, **23**, 47–64

Yu, G. H. and Kramer, P. J. (1969). 'Radial Transport of Ions in Roots', *Plant Physiol. (Lancaster)*, **44**, 1095–1100

Zholkevich, V. N. and Koretskaya, T. F. (1959). 'Metabolism of Pumpkin Roots During Drought', *Fiziologiya Rast.*, **6**, 690–700

Ziegler, H. (1956). 'Untersuchungen über die Leitung und Sekretion der Assimilate', *Planta (Berl.)*, **47**, 447–500

242   References

Ziegler, H. (1958). 'Über die Utmung und den Stofftransport in den Isolatierten Leitbundeln der Blattstiele von *Heracleum mantegazzianum* Somm. et Lev.', *Planta (Berl.)*, **51**, 186–200

Ziegler, H. (1960). 'Untersuchungen über die Feinstruktur des Phloems. I. Die Siebplatten bei *Heracleum mantegazzianum* Somm. et Lev.', *Planta (Berl.)*, **55**, 1–12

Ziegler, H. (1965). 'Use of Isotopes in the Study of Translocation in Rays', in *Isotopes and Radiation in Soil—Plant Nutrition Studies*. International Atomic Energy Agency, Vienna

Ziegler, H. and Kluge, M. (1962). 'Die Nucleinsäuren und ihre Bausteine in Siebrohrensaft von *Robinia pseudoacacia* L.', *Planta (Berl.)*, **58**, 144–153

Ziegler, H. and Vieweg, G. H. (1961). 'Der Experimentelle Nachweiss einer Massenströmung im Phloem von *Heracleum mantegazzianum* Somm. et Lev.', *Planta (Berl.)*, **56**, 402–408

Ziegler, H. and Ziegler, I. (1962). 'The Water Soluble Vitamins in the Sieve Tube Sap of Some Trees', *Flora (Jena)*, **152**, 257–278

Zimmermann, M. H. (1957a). 'Translocation of Organic Substances in Trees. I. The Nature of the Sugars in the Sieve Tube Exudate of Trees', *Plant Physiol. (Lancaster)*, **32**, 288–291

Zimmermann, M. H. (1957b). 'Translocation of Organic Substances in Trees. II. On the Translocation Mechanism in the Phloem of White Ash,' *Plant Physiol. (Lancaster)*, **32**, 399–404

Zimmermann, M. H. (1958). 'Translocation of Organic Substances in Trees. III. The Removal of Sugars from the Sieve Tubes of the White Ash (*Fraxinus americana* L.)', *Plant Physiol. (Lancaster)*, **33**, 213–217

Zimmermann, M. H. (1960a). 'Transport in the Phloem', *A. Rev. Plant Physiol.*, **11**, 167–190

Zimmermann, M. H. (1960b). 'Longitudinal and Tangential Movement within the Sieve Tube System of White Ash (*Fraxinus americana* L.)', *Beih. Schweiz. Forstv.*, **30**, 289–300

Zimmermann, M. H. (1961). 'The Removal of Substances from the Phloem', in *Recent Advances in Botany*. University of Toronto Press

Zimmermann, M. H. (1963). 'How Sap Moves in Trees', *Scientific Am.*, **208**, 132–142

Zimmermann, M. H. (1965). 'Water Movement in Stems of Tall Plants', in *The State and Movement of Water in Living Organisms* (Symposium No. XIX of the Society for Experimental Biology). Cambridge University Press

Zimmermann, M. H. (1969). 'Translocation Velocity and Specific Mass Transfer in Sieve Tubes of *Fraxinus americana* L.', *Planta (Berl.)*, **84**, 272–278

Zimmermann, M. H. and Brown, C. L. (1971). *Trees: Structure and Function*, with a chapter on 'The Irreversible Thermodynamics of Transport Processes' by M. T. Tyree. Springer, Berlin

Glossary of terms

Centrifugal transport	In stems, movement of solutes from the xylem to the extracambial tissues.
Centripetal transport	In stems, movement of solutes from the extracambial tissues to the xylem. In roots, from the external milieu to the xylem vessels.
Conduit (transport)	The cells through which solutes are transported.
'Leakage' (of solute)	Diffusional movement of solute molecules out of cells.
Loading (of solute)	Thermodynamically active transport process, occurring against a concentration or electrochemical potential gradient.
Orthostichy	Row of leaves which all lie in the same vertical plane on the stem.
Phyllotaxis (Phyllotactic configuration)	The arrangement of leaves on a stem.
Rate of translocation	The rate at which solutes are transported in a system, i.e. the mass transfer rate measured in the units, mass of solute transported/unit time, or specific mass transfer rate having the units, mass of solute transported/unit area of conducting tissue/unit time. Not to be confused with velocity of translocation.
Secretion (of solute)	Thermodynamically active transport of solutes against a concentration or electrochemical potential gradient.
Tangential transport	In stems, movement of solutes through the extracambial tissues in a direction parallel to a tangent drawn to the circumference of the stem.
Velocity of translocation	The speed at which molecules move through the transport conduits, measured in the units, distance/unit time.

List of Abbreviations

ABA	The naturally occurring growth retardant abscisic acid
ADP	Adenosine-5′-diphosphate
AMP	Adenosine-5′-monophosphate
ATP	Adenosine-5′-triphosphate
CTP	Cytidine-5′-triphosphate
2,4-D	2,4-Dichlorophenoxyacetic acid
DNA	Deoxyribonucleic acid
DNP	2,4-Dinitrophenol
EDTA	Ethylenediaminetetra-acetic acid
IAA	Indole-3-acetic acid
IAA-aspartate	Indoleacetyl aspartic acid
MCPA	4-Chloro-2-methylphenoxyacetic acid
NAA	1-Naphthalene acetic acid
OP	Osmotic potential
Paraquat	1,1′-Dimethyl-4, 4′-dipyridylium ion
Pichloram	4-Amino-3,5,6-trichloropicolinic acid
pmol	Picomole, 1×10^{-12} mole
p.p.m.	Parts per million
P-protein	Phloem protein, slime fibrils, plasmatic filaments, protein filaments
RNA	Ribonucleic acid
2,4,5-T	2,4,5-Trichlorophenoxyacetic acid
TCA	Trichloroacetic acid
TIBA	Tri-iodobenzoic acid
UDP-glucose	Uridine diphosphoglucose

Author index

245

Subject index